Sun in a Bottle?... Pie in the Sky!

L. J. Reinders

Sun in a Bottle?...
Pie in the Sky!

The Wishful Thinking of Nuclear Fusion Energy

 Springer

L. J. Reinders
Panningen, The Netherlands

ISBN 978-3-030-74733-6   ISBN 978-3-030-74734-3   (eBook)
https://doi.org/10.1007/978-3-030-74734-3

This Springer imprint is published by the registered company Springer Nature Switzerland AG
The registered company address is: Gewerbestrasse 11, 6330 Cham, Switzerland

Preface

Nuclear fusion is the process that powers the stars, including our own Sun. As soon as these stellar processes started to be understood (in the early 1920s), people began dreaming about harnessing their power both for the benefit and for the destruction of mankind. The development of the hydrogen bomb made the latter part of this dream come true. We now possess bombs that can destroy the Earth and all that is on it in a matter of hours or less. The other part of the dream, which concerns an inexhaustible clean source of energy that will save mankind from the horrors of climate change and pollution, has not yet become a reality.

For the past seventy years nuclear scientists and engineers have been trying to create this source of energy on Earth. So far in vain. From the early 1950s, promises have been made that its unlimited benefits will be available in at most two decades. The general media upped these promises with blazing headlines of the perceived breakthroughs that were achieved, while presenting the same old stories over and over again without taking the trouble to ask any critical questions. Examples are: "China's quest for clean, limitless energy heats up", "Speeding the development of fusion power to create unlimited energy on Earth", "'Star in a jar' could lead to limitless fusion energy". No other scientific or other enterprise I am aware of has ever been in need of so many 'breakthroughs' without making any real progress. Why is it taking so long?

The scientists themselves are partly to blame for this with silly statements galore in the literature and other places, such as the following one from the

1980s: "If the Martians were attacking, if money were no object and the military wanted a working fusion reactor by the year 2000, there is no question we could have it. By the year 2000 we could build such a BIG turkey." Such completely empty statements, only invoking hullabaloo, lack everything one expects of thoughtful scientists and are very reminiscent of statements made by certain politicians in the (fortunately short-lived) Trumpian age we just managed to survive. Science should steer clear of such talk and base its statements solely on science and scientific results. And the fact is that there are no such results or hardly any. The emperor has no clothes and never had any, and nobody wants to see it!

This book has grown out of a more extensive, more technical and more comprehensive version of the history of nuclear fusion in the last seventy years, called *The Fairy Tale of Nuclear Fusion*,[1] also recently published by *Springer Nature*. The goal of the present version is to present a more accessible, 'chatty' version of the fusion enterprise, aimed at the general public, exemplified by an intelligent eighteen-year old with a high-school education, who wants to look behind the screaming headlines about fusion's 'unlimited, abundant energy.' It tries to explain what has been and is now going on in fusion research and where it is likely to lead. When climate change came along, fusion scientists saw new chances and jumped early on this bandwagon, propagating fusion as the path to carbon-free, unlimited, clean energy and as the solution to the climate-change problems we are currently faced with. This book will shatter this prospect. There is no possible scenario in which fusion will make a sizable contribution to the energy mix in this century, let alone before or around 2050 as required by the Paris Climate agreement. Fusion will not make any positive contribution to the mitigation of climate change, nor will fusion energy be as clean and limitless as claimed by its proponents. If it ever becomes a reality, at the earliest in the course of the next century, electricity production from nuclear fusion will most likely be so expensive and so complex that it will never become economically viable.

There are no references in this book. If you want to know more about a certain topic or find out where it came from, please consult the book mentioned above, which contains extensive references. In the back of this book I will only give a general list of books on nuclear fusion and some related topics.

Panningen, The Netherlands   L. J. Reinders
March 2021

---

[1]L. J. Reinders, *The Fairy Tale of Nuclear Fusion* (Springer, 2021).

Contents

Acronyms and Abbreviations

AEC	Atomic Energy Commission
AERE	Atomic Energy Research Establishment
Alcator	Alto Campo Toro
APS	American Physical Society
ARIES	Advanced Reactor Innovation and Evaluation Study
ARPA-E	Advanced Research Projects Agency-Energy
ASDEX	Axially Symmetric Divertor Experiment
ASN	Autorité de Sûreté Nucléaire (French Nuclear Safety Authority)
BINP	Budker Institute of Nuclear Physics
CANDU	Canada Deuterium Uranium
CCFE	Culham Centre for Fusion Energy
CDA	Conceptional Design Activities (ITER)
CEA	Commissariat à l'énergie atomique et aux énergies alternatives
CFC	Carbon fibre composite
CFETR	China Fusion Engineering Test Reactor
CFS	Commonwealth Fusion Systems
CS	Central solenoid
DOE	Department of Energy (US)
DONES	DEMO Oriented Neutron Source
EAST	Experimental Advanced Superconducting Tokamak
ECRH	Electron Cyclotron Resonance Heating
EDA	Engineering Design Activities (ITER)
EFDA	European Fusion Development Agreement
ELM	Edge-localised mode
ETR	Engineering Test Reactor

Euratom	European Atomic Energy Community
FRC	Field-reversed configuration
HELIAS	Helical-Axis Advanced Stellarator
HLW	High-level waste
HTS	High-temperature superconductor
IAEA	International Atomic Energy Agency
ICRH	Ion Cyclotron Resonance Heating
IDCD	Imposed-Dynamo Current Drive
IFMIF	International Fusion Materials Irradiation Facility
IFRC	International Fusion Research Council
ILW	Intermediate-level waste
INFUSE	Innovation Network for Fusion Energy
INTOR	International Tokamak Reactor
IPP	Max Planck Institute for Plasma Physics
ISS	International Space Station
ITER	International Thermonuclear Experimental Reactor
ITER-FEAT	ITER Fusion Energy Advanced Tokomak
JAERI	Japan Atomic Energy Research Institute
JET	Joint European Torus
KSTAR	Korea Superconducting Tokamak Advanced Research
LCE	Lithium carbonate equivalent
LCFS	Last closed flux surface
LCOE	Levelized cost of energy or levelized cost of electricity
LH	Lower Hybrid
LLW	Low-level waste
LTS	Low-temperature superconductor
MAST	Mega Ampere Spherical Tokamak
MFTF	Mirror Fusion Test Facility
MIT	Massachusetts Institute of Technology
NBI	Neutral-beam Injection
NSST	Next Step Spherical Torus
NSTX	National Spherical Torus Experiment
ORNL	Oak Ridge National Laboratory
PF coils	Poloidal field coils
PFC	Plasma facing component
PKA	Primary knock-on atom
PLT	Princeton Large Torus
PPPL	Princeton Plasma Physics Laboratory
PSFC	Plasma Science and Fusion Center (MIT)
RAFM	Reduced activation ferritic/martensitic
RFP	Reversed Field Pinch
SIFFER	SIno-French Fusion Energy Center
SOL	Scrape-Off Layer
SPI	Shattered pellet injection

SST	Steady-state Superconducting Tokamak
ST	Spherical tokamak
START	Small Tight Aspect Ratio Tokamak
STEP	Spherical Tokamak for Energy Production
TBM	Test Blanket Module
TBR	Tritium breeding ratio
TCV	Tokamak à Configuration Variable
TF coils	Toroidal field coils
TFR	Tokamak de Fontenay-aux-Roses
TFTR	Tokamak Fusion Test Reactor
VLLW	Very-low-level waste
WEST	Tungsten (Wolfram) Environment in Steady-state Tokamak
ZETA	Zero Energy Thermonuclear Assembly

Units and Related Quantities

dpa   displacements per atom
eV   electronvolt (1.6×10^{-19} joule)
GW   gigawatt (1 billion (10^9) watts)
keV   kiloelectronvolt
MA   megaamperes
MeV   mega-electronvolt (1 million electronvolt)
MW   megawatt (1 million watts; 10^6 joule/second)
MWe   megawatt electric
ppm   parts per million
T   tesla, unit of magnetic field strength equal to 10,000 gauss
TW   terawatt (1 trillion (10^{12}) watts)

1

What is Nuclear Fusion?

The energy that reaches us from the Sun is the product of a process called nuclear fusion, the fusion of nuclei of atoms, the basic constituents of matter. Although the particular process that takes place in the Sun cannot be reproduced on Earth, the idea of generating vast amounts of energy from combining light elements into heavier ones has been a dream of mankind ever since the processes in the Sun and other stars were unravelled early in the twentieth century. Per kilogram of matter consumed the release of energy is about ten million times greater than in a typical chemical process like the burning of fossil fuels (coal, oil or gas). As will be discussed in this book, for the last seventy years a lot of effort has been put into trying to control nuclear fusion on Earth, in order to harvest this energy and solve the energy problems of mankind once and for all.

Apart from nuclear *fusion*, there is also the more commonly known process of nuclear *fission*, which concerns the splitting of nuclei of heavy elements into lighter ones. The best-known example is uranium, whereby the nucleus of this element is split into two smaller, more stable nuclei, while releasing at the same time a certain amount of energy. Per kilogram of matter the release of energy is, however, less than 10% of the output of fusion. Present-day nuclear power stations use this fission process to generate energy, something people also want to do with nuclear fusion by fusing light elements into heavier ones. In this book we will see that this is not such an easy matter.

To explain how all this works we first must know a little about nuclear physics, starting with the observation that all matter is built up of chemical

L. J. Reinders, *Sun in a Bottle?... Pie in the Sky!*, https://doi.org/10.1007/978-3-030-74734-3_1

Group →	1	2	3	4	5	6	7	8	9	10	11	12	13	14	15	16	17	18
↓Period																		
1	1 H																	2 He
2	3 Li	4 Be											5 B	6 C	7 N	8 O	9 F	10 Ne
3	11 Na	12 Mg											13 Al	14 Si	15 P	16 S	17 Cl	18 Ar
4	19 K	20 Ca	21 Sc	22 Ti	23 V	24 Cr	25 Mn	26 Fe	27 Co	28 Ni	29 Cu	30 Zn	31 Ga	32 Ge	33 As	34 Se	35 Br	36 Kr
5	37 Rb	38 Sr	39 Y	40 Zr	41 Nb	42 Mo	43 Tc	44 Ru	45 Rh	46 Pd	47 Ag	48 Cd	49 In	50 Sn	51 Sb	52 Te	53 I	54 Xe
6	55 Cs	56 Ba	*	72 Hf	73 Ta	74 W	75 Re	76 Os	77 Ir	78 Pt	79 Au	80 Hg	81 Tl	82 Pb	83 Bi	84 Po	85 At	86 Rn
7	87 Fr	88 Ra	**	104 Rf	105 Db	106 Sg	107 Bh	108 Hs	109 Mt	110 Ds	111 Rg	112 Cn	113 Nh	114 Fl	115 Mc	116 Lv	117 Ts	118 Og

*	57 La	58 Ce	59 Pr	60 Nd	61 Pm	62 Sm	63 Eu	64 Gd	65 Tb	66 Dy	67 Ho	68 Er	69 Tm	70 Yb	71 Lu
**	89 Ac	90 Th	91 Pa	92 U	93 Np	94 Pu	95 Am	96 Cm	97 Bk	98 Cf	99 Es	100 Fm	101 Md	102 No	103 Lr

Fig. 1.1 Periodic table of elements in the modern standard form with 18 columns

elements, such as carbon, iron, hydrogen, oxygen, etc. These elements are arranged in the Periodic Table of Elements (Fig. 1.1). The basis for this table was laid down in the late nineteenth century by the Russian chemist Dmitri Mendeleev (1834–1907). The smallest unit of an element (that still possesses the chemical properties of that element) is called an atom. Every solid, liquid, or gas is composed of atoms, in most cases neutral, i.e., uncharged atoms. Such an atom has a nucleus at its centre, which is orbited by negatively charged electrons.

Hydrogen (H), the lightest element occupies first place in the table, helium (He) is in second place, and so it goes all the way up to the heaviest element oganesson with number 118. The number of the element in the Periodic Table is, quite reasonably, called the *atomic number*, denoted by the symbol Z, and is equal to the number of electrons orbiting the nucleus. Electrons are negatively charged, and the total charge of all electrons equals the positive charge of the nucleus, making the atom as a whole neutral in charge. The positive charge of the nucleus is carried by particles, called protons, and the number of protons in the nucleus is equal to the number of electrons. The mass of a proton is about two thousand times the mass of an electron, which explains why most of the mass of an element is concentrated in its nucleus. So, the lightest element hydrogen has a single electron, and its nucleus is a single proton, while uranium, for instance, with atomic number 92 has 92 electrons orbiting a nucleus with 92 protons.

Apart from the charge-carrying protons, nuclei also contain a number of neutral particles, called neutrons, which have approximately the same mass as protons. Protons and neutrons are jointly called nucleons, and the number of nucleons in an atomic nucleus is called the *mass number*, denoted by the symbol A. The nuclei of all elements are built from such positively charged protons, supplemented with a number of neutrons. Since protons are positively charged, they don't play well together, and neutrons are needed to shield the protons from each other; else the nucleus would be unstable due to the repulsion of the charges. Protons repel each other and for a nucleus with more than one proton to be stable, one or several neutrons are needed. This is the reason why in nature no elements exist whose nuclei just consist of protons, apart from the trivial case of hydrogen.

The first 94 elements in the Periodic Table occur naturally on Earth and the remaining 24 are synthetic elements produced in the laboratory in nuclear reactions. Oganesson, for instance, was created in 2002, but only recognized as an element in 2015. It is not much of an element though as it falls apart very quickly and only five (possibly six) atoms of oganesson have ever been detected. Of the 94 natural elements, 83 are so-called primordial elements, meaning that they already existed before the Earth was formed. Of these, 80 are stable elements (1 through 82, i.e., hydrogen through lead, exclusive of 43 and 61, technetium and promethium, respectively), with three radioactive primordial elements (bismuth, thorium, and uranium). Uranium, which is unstable with a half-life of 4.6 billion years, is the heaviest of the primordial elements.

The Periodic Table does not show the whole story though, and there is a complication that cannot be seen from the table. The table only tells you the number of protons in the nucleus (or the number of electrons orbiting the nucleus) but does not say anything about the number of neutrons. Most elements come in a number of variants, which all have the same chemical properties, so the same number of electrons and protons (else it would be another element) but a varying number of neutrons. Such versions of an element are called *isotopes*: variants of a particular element that have the same chemical properties as that element but differ in the composition of their nucleus. The word comes from the Greek words *isos topos*, meaning *same place*, i.e., the same place in the Periodic Table. So, an element can appear in the guise of several isotopes.

Some such isotopes are stable, meaning that they will not fall apart, i.e., transmute (decay) into other elements or isotopes. For instance, the element tin, number 50 in the Periodic Table and denoted by *Sn*, is an isotope champion and holds the record with ten stable isotopes. They all have the same atomic number but differ in the mass number. Other isotopes are unstable

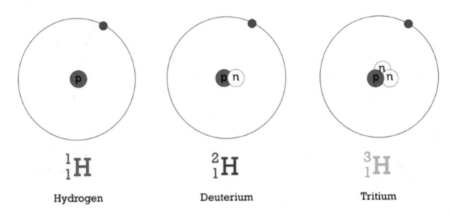

$$\begin{array}{ccc} {}^1_1\text{H} & {}^2_1\text{H} & {}^3_1\text{H} \\ \text{Hydrogen} & \text{Deuterium} & \text{Tritium} \end{array}$$

Fig. 1.2 Schematic representation of hydrogen and its isotopes

and will after some time decay, not necessarily into the element they were an isotope of in the first place. Such isotopes are said to be radioactive.

In Fig. 1.2 the three known isotopes of hydrogen are depicted. As can be seen in the figure the first isotope (deuterium) has one neutron in its nucleus apart from a proton, and the second isotope (tritium) has two neutrons. And, in general, the only difference between the isotopes of a certain element is the number of neutrons in their nuclei.

Of the isotopes of hydrogen shown in Fig. 1.2, ordinary hydrogen (^{1}H) and deuterium are stable while tritium is radioactive and decays into helium (^{3}He) plus an electron and a spooky particle, called a neutrino. Note the notation here with a left superscript denoting the mass number. To be complete a left subscript denoting the atomic number (place in the Periodic Table) should be added, as done in Fig. 1.2. This can be left out as the symbol for the element (H, He) already tells us what the atomic number is. The half-life of the decay of tritium is about 12 years, meaning that when you have a bunch of such tritium atoms and wait for 12 years half of them will have decayed. As we will see later, tritium is one of the main candidates for fusion fuel, together with deuterium. But it is not a very lucky choice, to put it mildly, as the radioactivity of tritium has extremely adverse, possibly even showstopping, consequences for fusion. Its radioactivity, although fairly mild, implies in the first place that all kind of precautions have to be taken, and secondly that it does not naturally occur on Earth and must be produced rather expensively in a special type of nuclear fission reactor.

Please note that the scale of the drawings in Fig. 1.2 does not reflect reality. Both the nuclei and the electrons occupy only a tiny part of the total atom, which mainly consists of 'empty' space filled with the electromagnetic fields generated by the charged nucleus and electrons. If the nuclei were as large

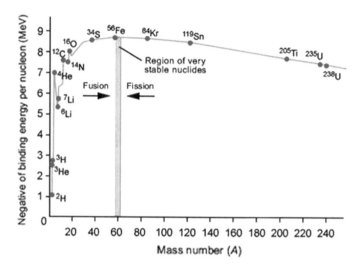

Fig. 1.3 Binding energy per nucleon plotted against the mass number

compared to the electron or the total atom as drawn in the figure, the electron orbit would not fit on the page.

How tightly bound are the nuclei of atoms and what is their mass? Nuclei consist of neutrons and protons, and one naively expects the mass of a nucleus to be equal to the sum of the masses of the protons and neutrons it consists of. That is indeed roughly the case, but not exactly, and this mass difference is precisely what the whole fuss is about. Many isotopes are lighter than expected from just adding up the masses of the nucleons in the nucleus or, saying it differently, a proton or neutron bound into a nucleus has slightly less mass than a free proton or neutron. This mass difference is called the *mass defect*. The energy equivalent to this mass defect, obtained from Einstein's famous formula $E=mc^2$, was released when the nucleus was formed from its constituent protons and neutrons.

The mass difference is therefore also the binding energy of the nucleus. In Fig. 1.3 the binding energy per nucleon, i.e., the total binding energy (or mass defect) divided by the number of nucleons in the nucleus, has been plotted against the mass number.

The figure shows that the binding energy per nucleon increases at first sharply with mass number and is largest for nuclei with mass number around 60, e.g., iron with atomic number 26 and mass number 56 (^{56}Fe), by far its most common isotope, is one of the most stable elements. Then the binding energy per nucleon slowly decreases down to mass number 240–250, the uranium isotopes. The consequence of this is that nuclei of elements heavier than iron can in principle yield energy by nuclear fission (in which

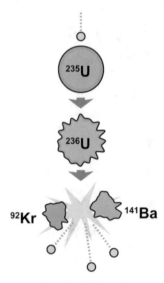

Fig. 1.4 Fission of a uranium-235 nucleus into barium and krypton

case they split into two more tightly bound nuclei), while elements lighter than iron can do this in principle by fusion (in which case they are fused into one more tightly bound nucleus). So, if a uranium-235 nucleus is split by bombarding it with neutrons, as happens in a nuclear fission reactor, many fission reactions are possible, but a typical one (with a neutron indicated by *n*) is:

$$n + {}^{235}U \rightarrow {}^{141}Ba + {}^{92}Kr + 3n$$

or in a figure format as represented in Fig. 1.4.

In this process the uranium nucleus first absorbs the neutron, forming the unstable uranium-236, which then splits into a barium and a krypton nucleus (with atomic numbers 56 and 36, respectively, adding up to uranium's 92) with the simultaneous emission of three extra neutrons. These neutrons can split further uranium nuclei and set off a chain reaction. The barium and krypton nuclei are bound much more tightly than uranium, resulting in the release of about 200 MeV or 200 million eV of energy, mostly carried away as kinetic energy by these 'daughter' nuclei, while the three neutrons that are released in the process carry away about 1 to 2 MeV each (4.8 MeV in total). An electronvolt (eV) is the amount of energy gained by a single electron when moving across an electric potential difference of one volt; it is equal to 1.6×10^{-19} J. This energy unit is the standard unit for the microworld of atoms and molecules, and will be used throughout this book. For comparison, 1 J is roughly equal to the kinetic energy of a tennis ball hitting the floor after falling from a height of 2 m. So, if a single tennis ball falling from a height of 2 m were enough to dent a packet of butter, to achieve the same by splitting uranium nuclei you would need the staggering number of 30 billion of such events (each event being worth about 3×10^{-11} J).

For fusion we have to be at the other end of Fig. 1.3, where the lightest elements at the very left of the figure are the most promising. In this respect it is especially important to observe that ^{4}He is particularly strongly bound, certainly compared to ^{2}H (or deuterium), so any fusion reaction that produces ^{4}He (for instance, by fusing two deuterium nuclei) will release a particularly large amount of energy.

Nuclear fission was discovered in 1938, about 20 years later than nuclear fusion, in Berlin by the German physicists Otto Hahn (1879–1968) and Fritz Strassman (1902–1980). Contrary to fusion, the discovery of nuclear fission immediately generated a lot of research, as it was soon realised that a self-sustaining nuclear chain reaction was possible, and a nuclear reactor could be built for harnessing the released energy. This was duly achieved in the first nuclear reactor, constructed in 1942 in Chicago as part of the Manhattan Project, America's effort to build an atomic bomb. Since then numerous nuclear fission reactors have been built and are still being built all over the world, providing a significant portion of mankind's energy needs. The price for this is, however, rather high, for many too high, in the form of waste products that due to their long-lasting radioactivity are difficult to dispose of, and in the form of accidents. Power from nuclear fission reactors rose rapidly in the second half of the twentieth century, stimulated in part by the oil crisis of the early 1970s and the supposedly scary thought of being dependent on the Arab countries for oil supply. Energy supply from nuclear fission reached a peak of a 17.6 percent share of globally generated electric power in 1996 and has declined ever since. As of August 2020, 440 reactors were operating in 30 countries. Its share of global electricity production has now declined to about 10%. The decline is reflected in the fact that more reactors are shut down than new constructions started (13 shutdowns versus just 5 construction starts in 2019).[1]

One of the reasons for this decline are accidents that have greatly diminished the appetite for fission reactors, especially in Western countries. The first large accident was the partial meltdown at the Three-Mile Island plant in Pennsylvania in 1979, followed in 1986 by the horrendous Chernobyl disaster. Especially the latter event sounded the death knell for power generation from nuclear fission. The last doubters were silenced by the Fukushima

---

[1]World Nuclear Association, 2020.

disaster in 2011, which like Chernobyl has left large areas of land contaminated with radioactive material. The Fukushima disaster led to panicky reactions of politicians all over the world, notably in Germany where plans were quickly drawn up to scrap all nuclear fission plants. It should be noted though that much of this panic was due to the way the disaster was reported in the press. Although nobody died in the nuclear accident with the Fukushima power plant, this aspect of the disaster got far more coverage in the press than the earthquake and subsequent tsunami which cost the lives of more than 20,000 people.

Although there is still quite a lot of activity in nuclear fission, as recalled above, its share of global electricity production will continue to fall. The public trust in nuclear fission as a power generation option has suffered a lethal blow and nobody seems to be able to turn the tide. This is unfortunate as many have argued that decarbonization of electricity generation will be a tough job, if not impossible, without nuclear fission plants. Whatever the rather fanatic anti-nuclear lobbyists are saying about nuclear fission power, it remains a fact that in the past (1980s–1990s) several countries, including France, Belgium, Switzerland and Sweden, managed to radically cut their greenhouse gas emissions by installing nuclear power. These countries now enjoy comparatively low carbon dioxide emissions, while the countries that have been installing renewable energy (solar and wind) in the last twenty years have hardly been able to cut their emissions and are still at a much higher level.

The bad reputation of fission power also has consequences for nuclear fusion, which likes to present itself as the safe nuclear option. That may well be the case, but it is a hazardous strategy as everything nuclear is viewed with suspicion by the public. Germany is a case in point, where the Green Party also opposes nuclear fusion as an energy option.

A further reason for the decline is the recent glut of cheap (shale) oil and gas coupled to a rapid increase in wind and solar energy, forcing nuclear fission power plants out of business, a trend that can only be reversed by slapping a substantial carbon tax on fossil fuels. A deeper reason for the decline is also that the technology of commercial nuclear fission reactors has stagnated. Nearly every nuclear fission power plant built in the last half century has been a light-water reactor, a design that in rare instances can indeed allow a meltdown and was aggressively marketed by the United States, which has now all but quit the field. Meltdown-proof, cheaper and more efficient designs, like very high-temperature reactors and other Generation IV nuclear reactors, have remained on the drawing boards for years, but are now being developed, mainly in China and Russia.

The reason why fission was developed in just a few years much earlier and much more extensively than fusion is in the first place that the technical realization of fusion is vastly more difficult than fission. The fundamental reason for this difficulty lies in the fact that fission can be achieved by firing neutral particles (neutrons) at nuclei, as shown in Fig. 1.4, while in the case of fusion, positively charged nuclei must be persuaded to fuse, which as we will see is a formidable task. Secondly, as regards fission, it was very soon realised that a bomb could be built on the basis of nuclear fission. For the latter the Manhattan Project was started up in the US in the early 1940s during World War II and fusion was put on a backburner for the duration of the war as the required temperatures, tens of millions of degrees, could only be achieved by a fission explosion. The technique of using a nuclear fission bomb as a 'matchstick' was eventually deployed in building the hydrogen or thermonuclear bomb. Thermonuclear bombs are still of this type. A pure fusion bomb (without the help of a fission bomb) has not yet seen the light, and hopefully never will. On the other hand, the fact that it hasn't, in spite of a colossal research effort, both by Western powers and by the Soviet Union, does not bode well either for fusion as an energy source. For if an uncontrolled release of fusion energy can apparently not be achieved without help from fission, how then can we have faith in controlled fusion ever being possible? Remember that in fission research it was the other way round. Scientists succeeded in keeping fission under control before a start was made with constructing an atomic bomb.

Fusion is conceptually a rather simple process, much simpler than fission and, more importantly, it is a great deal cleaner. The nuclei involved are much simpler, just a handful of nucleons compared to hundreds in the case of fission. Little radioactivity is released in the fusion process itself, and the radioactive waste it produces in a reactor is manageable, although a future nuclear fusion reactor is not as clean in this respect as many have wanted us to believe in the past. The radioactive waste problem in power generation from nuclear fusion is far from negligible, since, as we will see, the vast number of highly energetic neutrons released in the fusion processes will make much of the material of the reactor radioactive. In view of the general public's sensitivity to radioactivity it is paramount to be clear and transparent about this from the very beginning. A comparison with nuclear fission, which is indeed worse as regards long-lived radioactive waste, is not relevant in this respect, for something that is better than the perceived absolute evil is of course not necessarily good. Fusion may be well advised to avoid the comparison with fission as much as possible.

This brings us to the most important, and in the end perhaps decisive advantage of power generation from fusion over both fission and fossil fuels, namely that the primary fuel, the hydrogen isotope deuterium, can be obtained cheaply from water. This is one of the reasons that fusion is sometimes called "the ultimate energy source". In the water of the Earth's oceans one atom in every 6420 hydrogen atoms is deuterium, accounting for approximately 0.0156% (or 0.0312% on a mass basis, as a deuterium atom is twice the mass of ordinary hydrogen) of all naturally occurring hydrogen in the oceans. No mines are needed, no miners can get trapped, no transport of fuel to be burned in power stations, and a virtually inexhaustible supply. That is indeed true for deuterium, but in the currently preferred version of nuclear fusion, as we will see, tritium is also required, and this is in very short supply and moreover dangerously radioactive, making this claim of being the "ultimate energy source" a rather weak one.

As mentioned, fusion is technically much more demanding than fission, the root of the problem being that it cannot be induced by uncharged particles. The nuclei that must be brought together in a fusion process are all positively charged and, therefore, repel each other and want to be as far apart as possible. The larger the nuclei, the larger the charge and, since the repulsive force, called the Coulomb force, is proportional to the product of the charges of the nuclei, even for nuclei of moderately large atoms this repulsive force becomes prohibitive. This implies that, in order to have any chance of success, fusion fuels must be chosen from the lightest elements–hydrogen, helium, lithium, beryllium and boron. In spite of this small number of candidate elements, it still leaves us with more than 100 possible fusion reactions, of which those involving elements with only one charged particle in the nucleus (i.e., hydrogen and its family members) are the most promising.

Early progress in fusion devices for generating energy was also hampered by the fact that all fusion research, like fission research, not only for weapons development, but also for power generation, was kept secret until the late 1950s. The US, for instance, harboured fears that fusion reactors could be used as a neutron source to make bomb fuel. And indeed, a stimulus to most fusion research in the early days was the production of bomb-grade material for thermonuclear weapons and the fear of being left behind in their construction.

The nuclear powers at the time, the US, Britain and the Soviet Union, all started their own fusion research programmes after World War II and jealously guarded them from the outside world. They all employed essentially the same methods and techniques and encountered the same problems, yet none managed to construct a working fusion reactor. The fact that secrecy was lifted had undoubtedly to do with this lack of success. If any of the parties involved in fusion research in the early 1950s had made promising progress towards a working reactor, the secrecy would surely have become tighter still. This latter point is also borne out by the declassification guide jointly worked out by the British and Americans in 1957, which stated that "all information except that bearing on devices exhibiting a net power gain was to be opened." So, had there been any success with a working reactor, the information would have remained classified.

Since then it has gone up and down with fusion without any great success, although proponents would like us to believe otherwise. The Indian nuclear physicist Homi J. Bhabha (1909–1966), who chaired the 1955 First International Conference on the Peaceful Uses of Atomic Energy in Geneva, predicted at that conference that "a method will be found for liberating fusion energy in a controlled manner within the next two decades", i.e., by 1975. It has since become one of the clichés of nuclear fusion research that a commercial nuclear fusion reactor is ever only a few decades away. As the saying goes "nuclear fusion power is the energy of the future, and always will be". A former leader of the Tokamak Fusion Test Reactor (TFTR) at the Princeton Plasma Physics Laboratory recently stated, blaming insufficient funding, as is common practice among failing scientists and indeed people in most human endeavours, that "the goal of commercial fusion energy recedes 1 year per year", so as the Red Queen tells Alice in Lewis Carroll's "Through the Looking Glass," for fusion, too, 'it takes all the running you can do to stay in the same place'.

There is no other endeavour or project undertaken by mankind on which energy and money have been spent for close to a hundred years without any tangible results, and only a dim prospect of success in another fifty years or so. The reason must be that there is a lot at stake, or perceived to be: the promise of nuclear fusion power being an abundant, inexhaustible source of energy with little or no side-effects, at any rate manageable side-effects, and "too cheap to meter". Although the latter argument no longer seems to hold, the rest is already too good to be true, and if true, not something you would like to miss out on. No wonder that large teams of scientists in many countries are still working hard to try to solve the colossal scientific and technical problems involved in nuclear fusion. It would be a major achievement if in 25–40 years from now a working reactor for demonstration purposes were to become available, meaning a reactor which demonstrates that it is possible to build reactors that consume less energy than is needed to run them. This book intends to show that the chances for this to happen are very slim indeed.

2

Stellar Processes and Quantum Mechanics

As already mentioned, nuclear fusion is the source of energy in the Sun and in this chapter we will say a little more about this, starting with a 1920 article in the journal *Science* by the English astronomer, physicist and mathematician Arthur Eddington (1882–1944). He proposed that large amounts of energy released by fusing small nuclei might provide the energy source that powers the stars, although he had no idea yet how this would work. This is all the more remarkable as he was not even sure about the actual structure of atoms and the relationship between the various elements in the Periodic Table. His proposal predated the advent of quantum mechanics and a possible mechanism for such fusion was unknown. The only forces known to Eddington were electromagnetism and gravity. Gravitational contraction, i.e., the contraction of an astronomical object due to the influence of its own mass, drawing matter inwards towards the centre of gravity, was known to be responsible for star formation, and it had been estimated that if this were the source of the Sun's radiation it could only shine for about 20 million years. The Sun's surface would need to drop by about 35 m per year to provide enough energy from such gravitational contraction. So, there was a problem there, as around the same time geologists had shown that the Earth was at least two billion years old. In actual fact both the Sun and the Earth are as old as 4.6 billion years.

In his paper Eddington says: "*A star is drawing on some vast reservoir of energy by means unknown to us. This reservoir can scarcely be other than the sub-atomic energy which, it is known, exists abundantly in all matter; we sometimes*

© The Author(s), under exclusive license to Springer Nature
Switzerland AG 2021
L. J. Reinders, *Sun in a Bottle?... Pie in the Sky!*,
https://doi.org/10.1007/978-3-030-74734-3_2

dream that man will one day learn how to release it and use it for his service. The store is well-nigh inexhaustible, if only it could be tapped. There is sufficient in the sun to maintain its output of heat for 15 billion years." Now a century after Eddington wrote these words, we are still dreaming of tapping this source of energy, and it looks as if we are indeed getting a little closer, but the day of actually realizing this dream may still be far in the future or, more likely, remain elusive forever.

Although no real explanation of these stellar processes was possible before the advent of quantum mechanics, Eddington goes on to say that he believes *"that some portion of this sub-atomic energy is actually being set free in the stars."* He based his belief on experiments carried out by the English chemist and physicist Francis William Aston (1877–1945), which in Eddington's mind had conclusively shown that *"all elements are constituted out of hydrogen atoms bound together with negative electrons."* The structure of an atomic nucleus was not yet known at the time. It was thought to consist of an assembly of protons and electrons, the only elementary particles then known.

In his paper Eddington went on to state that, more importantly, Aston's precise measurements had also shown that *"the mass of a helium atom is less than the sum of the masses of the 4 hydrogen atoms which enter in it. (…) There is a loss of mass in the synthesis amounting to about 1 part in 120. (…) Now mass cannot be annihilated, and the deficit can only represent the mass of the electrical energy set free in the transmutation."* In the previous chapter we called this deficit the mass defect.

The equivalence of mass and energy, embodied in the formula $E = mc^2$, was proposed by Einstein in 1905. Because of the factor c^2 in this formula, with c the speed of light in vacuum being equal to about 300,000 km/s, even a minuscule amount of mass is equivalent to an awesome amount of energy. Where the chemical reaction of burning 100 grammes of coal would release 1 million joules of energy, the mass of these 100 grammes would according to Einstein's formula actually be equivalent to 10 million billion joules of energy, if only we knew how to get that energy out.

Einstein's formula was of course well-known to Eddington and he used it to calculate the amount of energy released when helium is made out of hydrogen. He concludes: *"If 5% of a star's mass consists initially of hydrogen atoms,[1] which are gradually being combined to form more complex elements, the total heat liberated will more than suffice for our demands, and we need look no further for the source of a star's energy."* And *"If, indeed, the sub-atomic energy in the stars is being freely used to maintain their great furnaces, it seems to bring*

---

[1]We now know that it is actually around 75%.

a little nearer to fulfilment our dream of controlling this latent power for the well-being of the human race – or for its suicide." The final part of his paper, which as we now know contained much truth, is a rather lengthy apology on his part for having in the eyes of many gone over to speculation.

How does this energy production in stars come about? A star starts off as an interstellar cloud of gas, mainly consisting of hydrogen, which begins to collapse under the influence of gravity as soon as it is massive enough for the gravitational forces to be stronger than the internal pressure in the gas. The star becomes ever denser and hotter until at some point the temperature becomes so high that hydrogen nuclei start to fuse into helium, according to the process to be described in greater detail in the next chapter, and energy is radiated off into space to warm planets like the Earth. In the star it increases the temperature still further and forces the gas to expand, countering the inward gravitational contraction. This results in an equilibrium whereby the star is held together by its own gravity and the internal gas pressure prevents it from collapsing further. This process continues until all the hydrogen has been burned away, after which a further contraction follows and other fusion processes take over, but it is in the first place the gravitational attraction that gets the process going.

This is also the way it works in the Sun. Being more than 300,000 times more massive than the Earth, the Sun can generate sufficiently large gravitational forces. It will be clear that gravity on Earth is (fortunately) much too weak to bring about such a contraction. If we want to establish fusion, the gas has to be compressed in another way. That this might be extremely difficult can be surmised from the fact that there are two forces competing here, the inward compression (in stars by gravity) and the outward pressure by the gas heating up. When a star that is powered by burning hydrogen into helium, like the Sun, has exhausted all its hydrogen, its core will become denser and hotter while its outer layers expand, eventually transforming the Sun into what is called a red giant. It will become so large that it engulfs the current orbits of Mercury and Venus, rendering Earth uninhabitable. But this will not happen for another five billion years or so. After this, it will shed its outer layers and become a dense type of star known as a white dwarf. It will be very dense with a volume comparable to Earth and no longer produce energy by fusion, but still glow and give off heat from its stored thermal energy from previous fusion reactions.

A star contains a hot burning core in which the fusion processes take place; the burning does not occur throughout the star. Eddington calculated that the temperature at the Sun's core would have to be about 40 million degrees, which is two to three times as hot as the currently accepted value of about 15 million degrees.

But there was another rather pressing problem. How could four protons (nuclei of hydrogen), all positively charged, come together to form the nucleus of a helium atom? The protons would repel each other and there was no way in Eddington's time to see how this repulsive Coulomb force could be overcome. Moreover, according to the classical laws of physics, the temperatures existing in the Sun were far too low for such fusion processes to take place. To find an explanation for this puzzle, quantum mechanics was needed, a new theory that was developed in the 1920s, mainly in Germany, in which utterly impossible things are allowed to happen.

It was the Russian physicist George Gamow (1904–1968) who, in 1928, while on leave in Göttingen from the Leningrad Physico-Technical Institute, added a vital ingredient to the solution of the puzzle by introducing the mathematical basis for what is known as *quantum tunnelling*. He saw that all the quantum physicists in Göttingen were beavering away at trying to understand the quantum mechanics of atoms and molecules, and instead of joining this crowded fray, he decided to have a look at what quantum theory could do for the atomic nucleus. In the library he had come across an article describing an experiment on the scattering of α-particles (an alternative name for nuclei of helium atoms) on uranium. From the scattering pattern it was clear that the α-particles were unable to penetrate the uranium nucleus. In itself not a strange result when one realizes the strong repulsive Coulomb forces between the positively charged α-particles and the positively charged uranium nucleus.

But, so Gamow asked himself, if that is the case how then is it possible that uranium, being a radioactive element, does itself actually emit α-particles which have about half the energy of the α-particles used to bombard the uranium nuclei? Apparently, a barrier prohibits the α-particles of the radioactive decay from getting out of the uranium nucleus for a rather long time. So how then can they get out? Gamow immediately realized what the answer should be. In quantum mechanics, unlike classical Newtonian mechanics, there are no impenetrable barriers, and there is a non-zero probability for a particle, that also can be described by a wave, to tunnel through a barrier (Fig. 2.1). The figure shows the potential barrier encountered by the particle, similar to a golf ball that has to get into a hole on the top of a little hill. The golf ball must scale the hillside before it can drop into the hole. Not so for a quantum mechanical particle, which can be described by an oscillating wave and has a non-zero probability to tunnel through the barrier and get into the hole, even if it does not have the energy to climb to the top. When encountering the barrier, the wave doesn't end abruptly. Instead, it continues

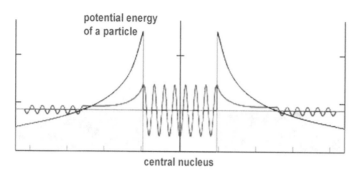

potential energy
of a particle

central nucleus

Fig. 2.1 A schematic picture of quantum tunnelling

inside and on the other side of the barrier, albeit with a smaller amplitude. Tunnelling gives a nonzero probability of finding a particle on the other side of a barrier. So, it doesn't behave like a golf ball at all!

A year later Robert d'Escourt Atkinson (1898–1982) and Fritz Houtermans (1903–1966) applied Gamow's tunnelling to provide the first calculation of the rate of nuclear fusion in stars. Their paper can be seen as the start of thermonuclear fusion energy research. The word 'thermonuclear' indicates the extremely high temperatures required in such nuclear processes. They give the particles a large enough thermal energy to overcome the Coulomb repulsion. Atkinson and Houtermans showed that, because of Gamow's tunnelling, fusion can occur at lower energies than previously believed (Eddington's 40 million degrees in the Sun's core) and that, in the fusing of light nuclei, energy could be created in accordance with Einstein's formula of mass-energy equivalence.

The energy released in the fusion of light elements is due to the interplay of two opposing forces. Protons are positively charged and repel each other due to the Coulomb force, but when they come very close together, due to their high thermal energy, quantum tunnelling allows the attractive nuclear force to overcome the repulsion of the Coulomb force and attract the nuclei further towards each other. This nuclear force is short-range, i.e., it is only felt when the nuclei are very close to each other (less than 10^{-15} m, a distance comparable to their size). Light nuclei are sufficiently small and have few protons. This allows them to come close enough to feel the attractive nuclear force. But to make this happen, extremely high temperatures and pressures are needed.

3

Nuclear Fusion of Light Elements

At the centre of the Sun, where the fusion takes place that eventually provides us on Earth with energy and light, the temperature is around fifteen million degrees. At this temperature, the electrons of the hydrogen atoms that make up about 75% of the Sun's mass have been stripped away. The resulting positively charged nuclei (protons) and unbound negative electrons move around with extremely high velocities in a very dense gaseous state (ten times the density of lead). This dense gaseous state is called a plasma and will be discussed in greater detail in the next chapter. The energy in the Sun is created by fusing protons in the plasma into helium. The process, called the "proton-proton" chain, involves three steps and was identified in 1939 by the German-American physicist Hans Bethe (1906–2005).

The first step involves the exceedingly rare process of the fusion of two protons. On average it takes a billion years for a proton to fuse with another proton and the proton-proton fusion processes taking place in 1 m^3 volume of the Sun produce just 30 W of heat, less than the heat on average given off by a human body. If the fusion rate were much higher, the Sun would burn up rather quickly and it would soon be over for us here on Earth, so we would not be able to arrogantly comment on the inefficiency of the Sun's fusion process. Fortunately, there is a huge amount of hydrogen present in the Sun, and at the Sun's temperature, this means that hydrogen-hydrogen fusion can take place frequently enough to keep the Sun burning for our benefit for a very long time. In spite of the rarity of the process, the Sun fuses in its core a staggering 600 billion kilograms of hydrogen every second giving 596 billion

L. J. Reinders, *Sun in a Bottle?... Pie in the Sky!*, https://doi.org/10.1007/978-3-030-74734-3_3

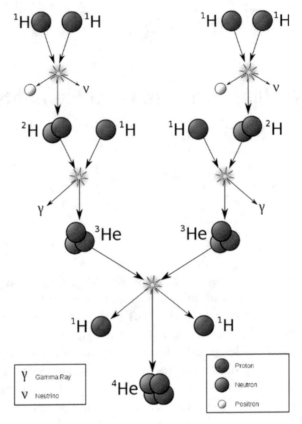

Fig. 3.1 The proton-proton chain reaction dominant in stars with the size of the Sun or smaller

kilograms of helium, and releasing the difference of 4 billion kilograms as energy. It has been fusing hydrogen into helium for around 4.6 billion years and has enough left to continue this process for another 4.6 billion years.

The precise process of the hydrogen fusion in the Sun is shown in Fig. 3.1. Every now and again (with a probability of once in a billion years) two hydrogen nuclei, i.e., two protons, come close enough together for the top process in the figure to take place. Two hydrogen nuclei fuse into deuterium. In this process one proton changes into a neutron by a rare process whereby a proton emits a positron (positively charged electron) and converts into a neutron and a spooky neutrino.

In the second stage of the process the deuterium nuclei that resulted from the proton-proton fusion will in turn capture another proton and form a helium-3 nucleus, which consists of two protons and a neutron. The chain then continues by two such helium-3 nuclei fusing into a helium-4 nucleus (which compared to helium-3 has one more neutron in its nucleus) and two protons. Helium-4 (^{4}He) is much more stable than helium-3 (^{3}He) and is the dominant isotope in naturally occurring helium. Thus, the end result in the Sun's fusion process is that four protons fuse into one helium-4 nucleus (α-particle), with the release of a certain amount of energy carried away by positrons, neutrinos and γ rays. This energy is what we are after. For a single fusion chain the release of energy amounts to the approximately 0.7% mass difference between the helium-4 nucleus and the four free-moving protons which formed the starting point of the process, or 26 MeV (26 million eV). This is still a very small amount of energy, but compared to the most energetic chemical reactions in which only a few electronvolts are released, nuclear reactions involve millions of times more energy. So, if a large number of fusion reactions takes place in a very short time or a self-sustaining chain reaction can be achieved, the total energy output will be enormous, as witnessed by the Sun and other stars.

The proton-proton chain reaction of Fig. 3.1 is a rather elaborate process and not suitable for reproduction on Earth because of the very low probability of a proton fusing with another proton. It is simply not possible to create a sufficiently large amount of hydrogen plasma on Earth to make this process work. Gravity on Earth is much too weak and the reaction rate of the proton-proton fusion reaction into deuterium is much too low. We just have not got the time.

A simpler, much faster method is to start directly with the hydrogen isotope deuterium, which is usually, for historic reasons, denoted by the symbol D (instead of ^{2}H which would be the proper designation). There is then no need to convert a proton into a neutron. Deuterium has a neutron in its nucleus and two deuterium nuclei already contain the necessary numbers of neutrons and protons from the start, resulting in a reaction which is 10^{24} times more probable than the Sun's proton-proton process. Roughly half of the time, the fusion reaction of two deuterium nuclei produces a ^{3}He nucleus and a neutron, with energies of 0.82 and 2.45 MeV, respectively. The other half of the time they fuse into a tritium nucleus (^{3}H) and a proton. This has been shown in the equation below, where for convenience the nuclear composition of the various particles has also been indicated. The third possibility (not indicated) of two deuterium nuclei fusing directly into ^{4}He is very rare.

$$D(np) + D(np) \rightarrow {}^3He(npp) \ (0.82 \text{ MeV}) + n(2.45 \text{ MeV})$$
$$\text{(half of the time)}$$

$$\rightarrow {}^{3}\text{H}(\text{nnp})\ (1.01\ \text{MeV}) + p(3.03\ \text{MeV})$$
$$\text{(half of the time)}$$

Even more convenient, deuterium is a stable isotope and abundantly present on Earth, where about 0.0156% of the water in the oceans is deuterated water HDO or D_2O (heavy water).

The interest in the other hydrogen isotope, tritium (${}^{3}\text{H}$ or more commonly T), for nuclear fusion lies in the fact that the least difficult and energetically most profitable fusion reaction is the one between deuterium and tritium:

$$D(\text{np}) + T(\text{nnp}) \rightarrow {}^{4}\text{He}\ (\text{nnpp})\ (3.5\ \text{MeV}) + n(14.1\ \text{MeV})$$

The peak fusion rate of this process is 10^{25} times higher than that of the proton-proton process in the Sun, with the peak occurring at about 100 million degrees. Comparing this with the Sun's process whereby a proton fuses on average in about 10^{16} s, it implies that, if the deuterium-tritium (D-T) process were available in the Sun, the latter would have been gone in a flash while releasing a staggering amount of energy, and there would have been no humans, at any rate not on this planet, to tell the tale. As can be seen by comparison with the deuterium-deuterium (D-D) reaction above, the fusion products in the D-D reactions (${}^{3}\text{He}$ and neutrons or tritium and protons) are 5–6 times less energetic than the products (${}^{4}\text{He}$ and neutrons) of D-T fusion. The energy of the neutrons is 2.45 MeV compared to 14 MeV for the neutrons released in D-T fusion, and this can be converted into heat, making the fusion reaction of deuterium and tritium the most energetically favourable.

Tritium is radioactive and decays naturally into ${}^{3}\text{He}$ with a half-life of about 12 years. On Earth it is present as a trace element in air and in water; about a billionth of a billionth (10^{-16} percent) of natural hydrogen is tritium. The most common forms of tritium are tritium gas (HT) and tritium oxide, also called tritiated water, in analogy with deuterated water. In tritiated water, a tritium atom replaces one of the hydrogen atoms, so its chemical form is HTO, like HDO, rather than H_2O for ordinary water.

Because little tritium is naturally present, it must be produced artificially for use on a practical scale. Tritium can be made in nuclear fission reactors that have been specially designed to optimize the generation of tritium and other special nuclear materials. It is produced by neutron absorption by a lithium-6 atom. This can be done by bombarding lithium-6 (7% of natural lithium is lithium-6, the more common form being lithium-7) with neutrons in a nuclear reactor, where lithium is added to the cooling water in order to regulate the acidity (pH) of the water. The hope is that in future the necessary tritium for nuclear fusion reactors can be generated by the same reaction in the fusion reactor itself by embedding the reactor core in a lithium blanket which will absorb neutrons from the fusion process and produce tritium:

$$^{6}\text{Li} + n \rightarrow {}^{4}\text{He}(2.05 \text{ MeV}) + \text{T} (2.75 \text{ MeV})$$

This is an energy producing reaction, no energy is needed to get the reaction going. If the more common form of lithium, ^{7}Li, is used instead of ^{6}Li, tritium will also be formed, in combination with ^{4}He and an extra neutron (since the ^{7}Li nucleus has one neutron more than ^{6}Li), but in that case 2.5 MeV of energy is needed for the reaction to occur.

The D-T fusion reaction presented above may be the most favourable energetically, but in other respects it is a terrible fusion reaction. Apart from complications arising from the radioactivity of tritium, such as remote handling and other precautions, the D-T cycle has two further principal disadvantages: (i) certain sensitive parts of the reactor require extensive shielding against the very high-energy neutrons released in the fusion processes, while other parts will be damaged and activated (another word for being made radioactive) by these neutrons, severely shortening their lifetime and needing frequent replacement, during which time the reactor will necessarily be shut down; (ii) the need, mentioned above, to breed tritium in the reactor itself, which results in considerable extra complexity and cost, and requires space for a lithium blanket.

The large energy release in D-T fusion results in minimal fuel consumption: the deuterium contained in 1 L of sea water (about 30 mg) and used in D-T reactions will produce as much energy as burning 250 L of gasoline. The burning of gasoline, a carbohydrate, is a chemical process driven by the outer electrons orbiting the nucleus in an atom. Fossil fuels are either carbohydrates or hydrocarbons, like natural gas. Carbohydrates consist of carbon, hydrogen and oxygen. Hydrocarbons are still simpler, just carbon and hydrogen: a certain number of hydrogen atoms chemically bound with carbon atoms into a molecule. The simplest molecule of this kind is methane CH_4, the next simplest is called ethane and has 2 carbon atoms and 6 hydrogen atoms (C_2H_6), etc. In the burning process the chemical bonds, binding the atoms into molecules, are broken, releasing energies of the order of 10 eV. This is infinitesimal compared to fusion reactions, which fuse the nuclei of atoms and involve energies of tens of MeV, hence a *million* times more. The enormous potential of nuclear fusion as an energy source can also be seen from the

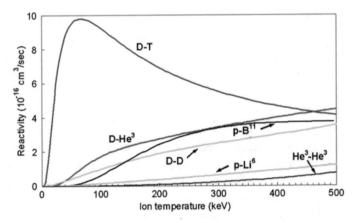

Fig. 3.2 Reactivity or reaction rate (the probability of a fusion reaction) versus ion temperature for some possible fusion reactions discussed in the text

following comparison. To generate 1 gigawatt of energy for one year (which is the typical need for a large industrial city) a conventional power station would have to burn 2.5 million tons of coal (causing an environmental pollution of 6 million tons of CO_2). A nuclear fission plant would need 150 tons of uranium (which would produce several tons of highly radioactive waste). A nuclear fusion power station would require just one ton of lithium (to breed the quantity of tritium needed, although it still has to be shown that this can be made to work), plus 5 million litres of water (for cooling), and would of course not release any greenhouse gases into the atmosphere.

There is, however, a huge snag in all of this: the temperature of the deuterium plasma in which the fusion must take place has to be about 150–200 million degrees, more than ten times as hot as the core of the Sun, while the D-T reaction is most probable at about 120 million degrees, in addition to a few other conditions. This temperature is needed to overcome the repulsion between the charged nuclei of tritium and deuterium. In the rest of this book we will see how much progress has been made towards this goal.

Other light elements are also candidates for fusion and in principle there are some 100 possible fusion reactions, which are, however, vastly less probable. The fusion rates for a number of them have been reproduced in Fig. 3.2, which clearly shows that D-T is by far the most favourable combination. In the figure the temperature has been expressed in keV. The concept of temperature will be discussed in the next chapter; for the moment it is sufficient to know that an energy of 1 eV is equivalent to about 12,000 degrees, so 1 keV corresponds to 12 million degrees.

The most promising reactions would be those that do not produce any (high-energy) neutrons (so-called aneutronic reactions, as activation of the reactor structure would be avoided) or reactions that only produce charged particles (which can be influenced and contained by a magnetic field). Some possibilities are:

Deuterium − Helium-3 $D + {}^3He \rightarrow {}^4He$ (3.6 MeV) $+ p$(14.7 MeV)

Deuterium − Lithium-6 $D + {}^6Li \rightarrow 2{}^4He$ (22.4 MeV)

Proton − Lithium-6 $p + {}^6Li \rightarrow {}^4He$ (1.7 MeV) $+ {}^3He$ (2.21 MeV)

Helium-3 − Lithium-6 ${}^3He + {}^6Li \rightarrow 2{}^4He + p + $ 16.9 MeV

Helium-3 − Helium-3 ${}^3He + {}^3He \rightarrow {}^4He + 2p + $ 12.86 MeV

Proton − Lithium-7 $p + {}^7Li \rightarrow 2{}^4He + $ 17.2 MeV

Proton − Boron-11 $p + {}^{11}B \rightarrow 3{}^4He + $ 8.7 MeV

Proton − Nitrogen $p + {}^{15}N \rightarrow {}^{12}C + {}^4He + $ 5.0 MeV

Tritium − Helium-3 $T + {}^3He \rightarrow {}^4He + p + n + $ 12.1 MeV (51%)

$\rightarrow {}^4He$ (4.8 MeV) $+ D$ (9.5 MeV) (43%)

$\rightarrow {}^5He$ (2.4 MeV) $+ p$ (11.9 MeV) (6%)

Of these the most commonly considered are the D-^{3}He, T-^{3}He, p-^{6}Li and p-^{11}B reactions.

The helium-helium reaction is part of the proton-proton cycle taking place in the Sun, as can be seen in Fig. 3.1. Since ^{3}He has two charged particles (protons) in its nucleus, helium-helium fusion is not a practical possibility on Earth. The other reactions also involve particles with several protons in the nucleus (Li and B even have 4 and 5 protons, respectively), which makes fusion virtually impossible due to the much stronger Coulomb repulsion.

Thus, the greatest difficulty with these alternative reactions is that fusion rates are much lower than for D-D and especially for D-T, as can be seen for some reactions in Fig. 3.2. This is the principal reason why these reactions have so far not played any role in practice, with the exception of proton-boron fusion, which is still actively pursued, although with little success. The prohibitive disadvantage seems to be that the temperature for fusion to occur must be roughly 1 billion degrees, a factor of ten higher than for the fusion of hydrogen and its isotopes. In the reaction a proton strikes a boron nucleus with an energy of about 500 keV, and this produces three alpha particles with an energy of 8.7 MeV. A major advantage of the reaction is that the reactor products are not reactive, as only helium is produced. Since there are no neutrons, no shielding and no blankets would be required; no tritium problems of any kind; no remote handling, while for ion temperatures above 200 keV the reaction rate would be even better than for D-D, but still lower than for D-T. Below 100 keV the reaction rate is almost negligible even compared to D-D. In the process only hydrogen and boron are used, and these are both plentiful on Earth, with ^{11}B being the main isotope of boron. All the energy, 8.7 MeV in total, is carried away by charged helium particles. They can be slowed down with electric fields and produce electricity directly. Quite a number of advantages, so no wonder it has always drawn a fair amount of attention, although it has been concluded more than once in the literature that the p-^{11}B fusion rate is not large enough to achieve fusion because of unavoidable losses from ion–electron scattering.

A further complication is that, when considering reactions with elements with Z values above 2, such as lithium and boron, the number of protons and neutrons in the fusing nuclei is rather large and after fusion with a proton they can recombine in various ways, i.e., there are competing reactions apart from the ones given above, or the main fuel can react with products of the fusion reaction itself, e.g., ^{6}Li and a proton can fuse into ^{4}He and ^{3}He, after which the newly produced ^{3}He can react with any remaining ^{6}Li, regenerate the proton and leave in the end only ^{4}He particles. This possibility of multiple outcomes applies to most of the reactions given above. The more neutrons and protons there are available in the fusing nuclei, the more possible combinations in which they can recombine after the fusion process and the more possible outcomes.

Let us consider the aneutronic D-^{3}He reaction in a little more detail. It is the first reaction in the above list and is often held to be the answer to the disadvantages of D-T fuel mentioned above. When using a deuterium-helium mixture, some neutrons will still be produced by parasitic D-D reactions, i.e., deuterium fusing with other deuterium ions instead of with helium, but the number will be small and can be minimized by a careful choice of fuel mixture—in particular a so-called 'D-lean' fuel mix (with more than 75% ^{3}He). As can be seen from the D-^{3}He reaction above, the fusion energy is released in the form of charged particles (the protons have as much energy as the neutrons in the D-T reaction, but being charged, they can be kept under control by magnetic fields). This would make it possible to use some form of 'direct conversion' of fusion energy into electricity at higher efficiency than in a D-T reactor, where neutrons carry away 80% of the energy, which is deposited as heat. In the subsequent conversion of this heat into electricity, 60% is typically lost due to the constraints imposed by thermodynamics. That is why it has been suggested that a commercial fusion reactor based on the tokamak design and using D-^{3}He might be competitive with one using D-T. The physics conditions for D-^{3}He are, however, much tougher (in order to overcome the Coulomb repulsion, a plasma temperature of around 800 million degrees is needed for this fuel mixture) and the reaction rate is much lower than for D-T fusion (Fig. 3.2). These drawbacks might be compensated, so it is thought by some optimistic souls, by a reduction of the severe engineering difficulties for D-T reactors.

In addition, an obvious and serious limitation to the use of D-^{3}He is the lack of an adequate terrestrial source of ^{3}He. Lunar mining of ^{3}He has been proposed as a lunatic solution to this problem. It would require the development of an enormous industrial base on the Moon, which does not seem realistic at present or in the near future. A second possibility to produce helium would be its manufacture in a D-D fusion reactor. This would require not only D-^{3}He reactors but also D-D reactors, doubling the problems and losing the advantages, as tritium (from which the ^{3}He would be extracted) and neutrons would also be produced. Investigations by several scientists into the use of D-^{3}He fuel have led to the carefully worded conclusion that "it seems unjustified to claim that aneutronic fuels and direct conversion offer comparable promise to conventional D-T fuelled magnetic confinement systems". Bearing in mind that the promise of D-T fuelled confinement systems is also very slim indeed, this assessment seems near fatal.

In the following we will only be concerned with deuterium-deuterium and deuterium-tritium fusion. Although the latter process is the most favourable of the two, it has mostly been avoided because of the radioactivity of tritium. Only two experiments with tritium have been carried out to date, making the experience gained with this type of fusion fuel, which in the end is supposed to become the favoured fuel, very scant indeed. Most experiments work only with deuterium.

The challenge faced by nuclear fusion is to create a deuterium-tritium plasma of the required temperature, take care of its confinement so that a 'controlled fusion burn' can start, and tap the energy released in this process. A possible solution to tackle this problem was soon clear. Magnetic fields could be used to confine the plasma, making use of the fact that the plasma consists of charged particles, which apart from creating difficulties by their own magnetic and electric fields will also be influenced by external magnetic fields and can perhaps be shackled in a straitjacket of such fields. The further history of nuclear fusion is just that: various attempts to create such a straitjacket, or as they like to say, to put the Sun in a bottle.

This is obviously a huge engineering challenge. Before we elaborate on the history of this challenge, we will first discuss in more detail the concept of a plasma, as it seems obvious that before trying to shackle something, you should try to understand it, a maxim that is surprisingly not prevalent in the fusion community.

4

Plasma

As noted in the previous chapter, at the high temperatures required for D-T and other fusion reactions, the fusion fuel is in a gaseous state in which all atoms have been ionised. The electrons that normally orbit the atom's nucleus have been stripped away, and a mixture of positively charged nuclei (ions) and negatively charged electrons remains (although there are usually also some neutral particles present). This mixture is called a plasma and is sometimes called the fourth state or phase of matter. The presence of mostly charged particles makes a plasma hugely different from a solid, liquid or gas, which do not contain any or very few (partially) ionised atoms. In this chapter we will briefly discuss some of the properties of plasmas.

Depending on the circumstances, every form of matter can exist in any of four states: solid, liquid, gas or plasma, so any gas can be made liquid or solid or turned into a plasma. Heating a solid, it becomes liquid, heating it further it becomes a gas, and upon still further heating the gas will be ionised into a plasma. The thermal energy of the particles in the plasma becomes so great that the electric forces that ordinarily bind the electrons to atomic nuclei are overcome, causing the electrons to fly off. Instead of a hot gas (mainly) composed of electrically neutral atoms, a plasma is a swirling turbulent mix of oppositely charged particles—electrons and nuclei.

One of the chief characteristics of a plasma is that it is electrostatically neutral. That means that to a high degree of accuracy the negative charges of the electrons balance the positive charges of the ions, ensuring that in macroscopic volumes of plasma there is only a very small net electric charge.

© The Author(s), under exclusive license to Springer Nature Switzerland AG 2021
L. J. Reinders, *Sun in a Bottle?... Pie in the Sky!*,
https://doi.org/10.1007/978-3-030-74734-3_4

This is of course not surprising as we started from a gas and increased the temperature, which ionised the gas atoms, and all constituents of these atoms (electrons and nuclei) are still present.

This indicates that the physics of plasmas has one of its roots in the physics of gas discharges. A gas discharge is the release and transmission of electricity through a gas in an applied electric field that occurs due to the ionisation of the gas when an electric current flows through it. It goes back to the middle of the nineteenth century, when physicists started to investigate the conductivity of gases. Glass tubes were partially *evacuated* and had metal plates sealed at the ends. The metal plates were connected to a battery causing a current to flow between the plates. With decreasing gas pressure, luminous phenomena could be observed; the tube would begin to glow. The charged particles in the gas in the tube were accelerated by the applied voltage and, on colliding with the neutral gas molecules, stripped them of their electrons. In this collision process the ions released energy in the form of the emission of light of a characteristic frequency. The charged products of these processes (ions and electrons) were in turn accelerated and ionised other neutral gas molecules, causing the ionisation of the gas to grow as in an avalanche. The resulting mixture of ions, electrons and neutral atoms and molecules was called a plasma.

Another root of plasma physics lies in astrophysics, which is understandable as 99% of matter in the universe is in a plasma state. Early in the twentieth century astrophysicists became interested in cosmic magnetic fields, the magnetic fields of the Sun and other astronomical objects, the origin of sunspots, their magnetic fields, and suchlike. It was astrophysics that realised the importance of magnetic fields for plasma physics and discovered a number of important phenomena, such as plasma waves.

The fusion-oriented plasma physics was mainly formulated by plasma astrophysicists with, at least in the beginning, little participation by physicists working with plasmas in other fields. This has resulted in an underestimation of experimental difficulties. This underestimation is all too obvious in the history of nuclear fusion, among other things in the rash predictions that are commonly made.

Temperature

At this point it may be useful to briefly discuss the concept of temperature for a mass of particles in a gas. Temperature refers to the thermal motion (velocity) of particles, and a gas in thermal equilibrium, i.e., at a constant temperature in space and time, has particles of all velocities. This thermal motion of the particles determines the temperature of the gas and the "temperature" of an individual particle is its velocity. The higher the temperature of the gas, the more fast particles there are and the less slow particles. They obey a certain probability distribution, which in the simplest case for particle speeds in an idealised gas is a so-called normal distribution (Maxwell-Boltzmann distribution) in the form of a bell-shaped curve; a continuous probability distribution that is symmetric around an average value, which corresponds to the temperature of the gas. This is the temperature that is meant when we speak of the temperature of a gas or a plasma. It implies that the temperature is directly related to the energy of the particles. This energy, as we have seen in Chap. 1, is expressed in electronvolt (eV) and an energy of 1 eV is equivalent to a temperature of about 12,000 degrees.

An interesting phenomenon for a plasma consisting of ions and electrons is that it will have several temperatures at the same time, since the velocities of the ions and the electrons are different and obey different Maxwell-Boltzmann distributions with different average values. In due course they will become equal as they collide with each other and transfer energy, but a plasma does not generally live long enough for the two temperatures to equalize. For a certain period of time each species of particle is then in its own thermal equilibrium at its own temperature.

Another feature of temperature is that high temperature does not necessarily mean a lot of heat. The total amount of heat (the heat capacity) involved in a low-density gas, such as in a fluorescent tube, is not that much, not enough to cause burns or damage to the wall of the tube, as the amount of heat that would be transferred in a collision with the wall is rather small. The same is, for instance, true for the ash from a burning cigarette dropping on someone's hand. Although the temperature is high enough to cause a burn, it actually doesn't, as the total amount of heat is insufficient for this. For plasmas in nuclear fusion devices too, in spite of temperatures of millions of degrees the heating of the walls is not a serious problem.

Plasma on Earth and Elsewhere

Unlike the solid, liquid and gas phases, under normal conditions the plasma state does not exist freely on the Earth's surface. That does not mean though that all the other phases actually do exist under normal conditions for all matter. Helium gas, for instance, only becomes liquid at a few degrees above absolute zero (-273 °C), a temperature that does not occur under normal circumstances anywhere on the planet.

The plasma phase is different from the other three phases by the fact that on Earth there is no known type of matter which under normal conditions is in the plasma phase, as the temperatures required for forming a plasma are very high. On Earth we experience plasmas in short-lived phenomena such as the flash of a lightning bolt, the soft glow of the polar aurorae (e.g., Aurora Borealis) and ionospheric lighting, but also in the light of a fluorescent tube and the pixels of a plasma TV screen, which are etched with plasmas. These are not the fully ionised plasmas needed for fusion but are partially ionised, relatively low-temperature plasmas of about 4 eV (about 50,000°). Other applications are plasma etching and deposition to make semiconductors in electronic devices, and windows glazed with plasmas to transmit or reflect specific wavelengths of light.

Beyond the Earth's atmosphere the magnetosphere is a plasma system formed by the interaction of the Earth's magnetic field with the solar wind, the stream of charged particles released from the upper atmosphere of the Sun, the corona. Further off in the universe, the most common phase of matter is plasma. All matter of interstellar space is plasma, although at an extremely low density, while the interiors of all stars are actually made out of very dense plasma, to which we thank our existence. Without it, there would perhaps be a planet like Earth, but without the energy of the Sun, which comes to us every day, the emergence of life on this planet would not have been possible.

Physics of Plasmas

In physics terms a plasma is an ionised gas in which at least one of the electrons of an atom has been stripped free, leaving a positively charged ion. In any gas there is always some small degree of ionisation, but this does not mean that any ionised gas can be called a plasma. An atom in a gas can be ionised by suffering a collision of high enough energy that causes an electron to be knocked out. For nitrogen, the main component of air, about 14.5 eV is needed for this. We have seen earlier that 1 eV corresponds to about 12,000°, and at that temperature very few particles will have enough energy to knock an electron out of a nitrogen atom. Consequently, for air at room temperature the degree of ionisation is completely negligible. Other atoms, e.g., alkali metals, such as potassium and sodium, possess much lower ionisation energies, and plasmas may already be produced from them at temperatures of about 3,000°. To get appreciable ionisation in air, temperatures of around 100,000° are needed, and much higher still for a plasma to be formed. In order for the nuclei in a plasma to undergo fusion with a probability that can keep a fusion reaction going, the plasma temperature must be at least 100 million degrees. For the mere existence of a plasma such high temperatures are not needed. It is sufficient to have temperatures at which the atoms in the gas break apart into a gas of charged particles (electrons and nuclei) and to sustain this ionisation. This does not require extremely high temperatures. So, plasmas exist in an enormous temperature range from a few thousand degrees to 100 million degrees. It seems obvious that the properties of such plasmas can also differ enormously between such extremes.

For the purposes of this book it is sufficient to consider *a plasma as an electrically neutral medium of unbound positively and negatively charged particles, ions and electrons, that move about violently,* as the particles have a high thermal energy, i.e., high speed, bump into one other and feel each other's electric and magnetic fields. It is best compared to a boiling hot "soup" of negatively charged electrons and positively charged nuclei. The overall charge of a plasma is roughly zero, but the ionised gaseous substance forming the plasma becomes highly electrically conductive, and long-range electric and magnetic fields dominate its behaviour. In ordinary air the molecules are neutral in charge and do not experience any net electromagnetic forces. The molecule moves undisturbed until it collides with another molecule. In a plasma with its charged particles the situation is totally different. As the particles move around, local concentrations of positively or negatively charged particles can arise spontaneously, giving rise to electric fields. Motion of charges generates currents and thus magnetic fields, as a current always generates a magnetic field. These fields act on particles far away. Therefore, in a plasma many charged particles interact with each other by long range forces and various collective movements can occur, resulting in many kinds of instabilities and wave phenomena; waves of charged particles can propagate through the plasma without damping or amplification.

The electrical conductivity of hydrogen plasma, a plasma consisting of hydrogen nuclei (protons) and electrons, is about ten times higher than that of copper at normal temperature, and this conductivity is one of the most notable features of a plasma. The conductivity is so great that, within the main body of the plasma, externally applied electric fields are effectively cancelled by the currents they induce in the plasma. If two charged balls connected to a battery are inserted inside a neutral medium, an electric potential will arise between the balls. In a plasma with its mix of oppositely charged particles the balls will attract these particles. One ball will attract the positive ions and the other the negative electrons, such that the charges on the balls will be *shielded off* and no electric field will be present in the plasma between the two charged balls.

The fact that this hot gas consists of charged instead of neutral particles is both a blessing and a curse. It is a blessing as charged particles can be influenced and controlled, so it is hoped, by magnetic fields, so that the plasma can be confined long enough for fusion reactions to occur. It is a curse as the particles themselves also produce electric and magnetic fields, which makes modelling of plasma behaviour very complicated.

The interiors of stars like the Sun consist of a fully ionised hydrogen plasma with all the electrons stripped off the atoms; the bare nuclei and electrons swirl about and electric fields are rampant everywhere. The plasma in the Sun's core (which extends to about a quarter of the Sun's radius, i.e., 175,000 km) is very dense (ten times the density of lead). Its temperature is about 15 million degrees and the pressure at the centre is estimated at 265 billion bar. It is held together by the vast gravitational forces created by the Sun's mass, which is 300,000 times that of the Earth. Consequently, due to the relatively small mass of a planet like Earth, this force is far too weak to do the same job in terrestrial circumstances. Even in the Sun, gravitation is not a perfect confining force as a continuous stream of particles is escaping from the Sun's inner core. Perhaps that should be a warning: if even the colossal gravitational forces in the Sun cannot confine a plasma, how on earth can we be so arrogant as to think that we can do this with some puny magnetic field in a contraption here on Earth? It defines the next and main problem for any viable fusion reactor, to be discussed in the following chapters. How to maintain the plasma at the high temperature required and confine it in a vessel or any other device that might be able to do that job?

Secondly, as we will see, any plasma that will be created on Earth in nuclear fusion machines will have vastly different properties from the plasma in the Sun's interior. The Sun's core is in some sort of dynamic equilibrium with the gravitational forces acting on it. The gravitational pressure on the core is resisted by a gradual increase in the rate at which fusion occurs. This process speeds up over time as the core gradually becomes denser. It is estimated that the Sun has become 30% brighter in the last four and a half billion years. So the common references in nuclear fusion stories to the Sun or other stars, cf. the title of this book or the claim in one of the ITER brochures "a star is born", are far from true and, if anything, rather misleading and to some extent annoying as they present the wrong picture. We will never (be able to) create a "piece of the Sun" or "bottle the Sun" here on Earth. ITER (the International

Thermonuclear Experimental Reactor), currently under construction in the South of France and so far the high point in fusion research, will not be a star. All these references to the Sun are just a load of baloney.

5

Plasma in Nuclear Fusion Devices

Research in plasma physics really got going in the early 1950s when proposals were put forward to harness the power of nuclear fusion. At that early stage all proposals involved the application of magnetic fields in order to try to contain the plasma, as it seemed the only way to keep the plasma away from the wall of the vessel in which it was confined. A second approach to fusion, called *inertial confinement fusion*, has a quite independent history as it always was and still is closely related to weapons research. It will not be discussed in this book as it is not in the race for energy generation by fusion. In this and the following chapters we will only look at *magnetic confinement fusion*, the use of magnetic fields to contain the plasma.

Once the plasma has been achieved, the next problem is how to maintain it, prevent it from losing energy, and heat it to increase the likelihood of fusion. For making fusion a reality, we need enough plasma at a high enough temperature that holds its heat for long enough. The plasma, however, loses energy in various ways and energy is needed to heat it up. The energy required for all this must be less than the energy that is eventually produced, as there clearly would be little sense and interest in a power plant that produces less energy than needed to operate it. The energy balance must be positive and has come to be denoted by Q: the ratio of the energy gained to the energy lost. This is a seemingly simple quantity (just the ratio of what goes in to what comes out) but, as we will see in one of the future chapters, fusion scientists have managed to make it a very complicated and muddled business by defining all kinds of Q.

© The Author(s), under exclusive license to Springer Nature
Switzerland AG 2021
L. J. Reinders, *Sun in a Bottle?... Pie in the Sky!*,
https://doi.org/10.1007/978-3-030-74734-3_5

Lorentz Force and Particle Trajectory in a Magnetic Field

In fusion devices external magnetic fields are applied to confine the hot plasma in a certain region, preventing the particles from making contact with the vessel and the plasma from rapidly cooling down. In this section we will consider the motion of an individual charged particle in a magnetic field. This is eventually the level of detail needed to be able to make reliable models of a plasma in a magnetic field.

A charged particle in a magnetic field is subject to a force, called the Lorentz force, after the Dutch physicist Hendrik Anton Lorentz (1853–1928). The force has four main properties: (1) since a magnetic field has no grip on neutral particles, the force only acts on charged particles, in opposite directions for negative and positive charges (the force is proportional to the charge of the particle); (2) it only acts on moving particles (the force is also proportional to the velocity of the particle); (3) the force is proportional to the strength of the magnetic field; and (4) the action of the force is perpendicular to the velocity of the particle *and* to the direction of the magnetic field (Fig. 5.1).

In a uniform field with no additional forces, a charged particle will gyrate (describe a circle) around the magnetic field according to the component of its velocity that is perpendicular to the magnetic field, and drift parallel to

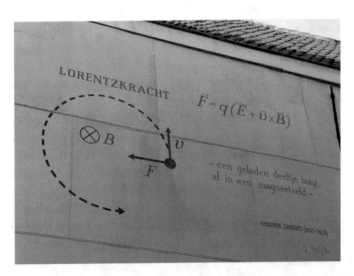

Fig. 5.1 The Lorentz force as depicted on a wall in Leiden (Netherlands) close to where Lorentz lived. The magnetic field *B* is perpendicular to the wall; the force *F* is directed inwards and the motion of the particle with velocity *v* is bent into a circle

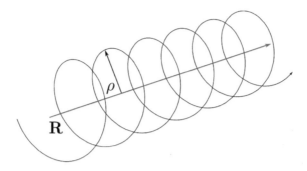

Fig. 5.2 Corkscrew motion of a charged particle in a magnetic field in the *R* direction (guiding centre); ρ, the radius of the particle's motion, is called the gyro-radius or Larmor radius

the field according to its initial parallel velocity. The result is that the particle describes a helical (corkscrew) motion around a guiding centre in the direction of the magnetic field (Fig. 5.2). This is the basic motion of any charged particle in a magnetically confined plasma. It can be treated as the superposition of the relatively fast circular motion around the guiding centre and the relatively slow parallel drift of this centre.

If the speed of the particle is constant, the force is directed inwards and has everywhere the same strength; the particle orbits are circles and the radius ρ is called the gyro-radius or Larmor radius, after Joseph Larmor (1857–1942), an Irish physicist and mathematician. In a fusion reactor a deuterium nucleus will typically have a gyro-radius of 1 cm, while the plasma itself will have a typical radius of about one metre. The nucleus will complete about 30 orbits per microsecond. An electron with the same energy has a much smaller radius than a deuterium nucleus. As it is much less massive it will move much faster, which at first sight would increase its radius. The Lorentz force (being proportional to the velocity) is, however, much stronger for an electron and the result is that the electron gyro-radius is about a factor of 60 smaller (the square root of the mass ratio of deuterium and an electron). The gyro-radius is proportional to the particle's mass and velocity, while it is inversely proportional to the particle's electric charge and the magnetic field strength. So, the larger the mass and/or velocity the bigger the circle, and the larger the charge and/or magnetic field the smaller the circle, which implies that increasing the field strength will squeeze the particles around the magnetic field lines.

If wires are wound in a helix around a vacuum tube and a current is sent through the wire, a magnetic field will be created in the tube. A charged particle travelling from one end of the tube to the other will then describe such helical motion around the magnetic field line, and be prevented from touching the walls of the tube, if it is not knocked off its track in a collision with any of the (neutral) particles left behind in the tube. With the proper configuration of magnetic fields this can also be done for a plasma, which is then suspended as it were in "mid air", in this case in a vacuum vessel. It can thus be prevented from making contact with the vessel wall and hopefully from losing too much energy. It was realised from the very beginning of nuclear fusion research that this is *the* essential and fundamental problem this research must solve, and its complexity was grossly underestimated. We will see in future chapters which "magnetic bottles" have been proposed for letting the particles go round and round. But let us now look at the conditions that are necessary for fusion to happen.

The Lawson Criterion

In the mid-1950s the English physicist John D. Lawson (1923–2008) derived the so-called Lawson criterion. It defines the conditions needed for a fusion reaction to reach *ignition*, i.e., the situation in which the heating of the plasma by the products of the fusion reactions is sufficient to maintain the temperature of the plasma against all losses without any external heating being required, in other words to keep the reaction running without outside help. In Chap. 3 we have seen that in D-T fusion α-particles (helium-4 nuclei) are produced in addition to the high-energy neutrons. The neutrons must in the end supply the electricity, while the α-particles must keep the plasma at the required temperature to keep the reaction going. In that connection we also speak of a *burning* plasma, in which the heat from the fusion reactions is almost or completely sufficient and produced for a sufficiently long time to maintain the temperature in the plasma. Once the fusion has been set in motion, enough energy is produced to keep the plasma at the required temperature.

Lawson calculated the requirements for this to happen. He came up with a dependence on three quantities: plasma temperature (T), discussed in Chap. 4, plasma density (n) and confinement time (τ). The Lawson criterion gives a minimum value for the product of these three quantities (called the triple product) required for the fusion reaction to become self-sustaining. The plasma density (usually denoted by n) is the number of fuel ions (nuclei of deuterium and tritium that can fuse) per cubic metre. The *energy confinement time*, usually denoted by the Greek letter τ with subscript E (τ_E), measures the rate at which the system loses energy to its environment, and is defined as the total amount of energy in the plasma divided by the power supplied to heat the plasma[1]:

$$\tau_E = \frac{\text{energy in plasma}}{\text{power supplied to heat plasma}}$$

It measures how well the magnetic field insulates the plasma, i.e., prevents it from losing energy. A plasma loses energy in various ways, some of it by conduction (the transfer of energy by collisions and movement of particles), by loss of particles or by heat radiation. Energy must be confined long enough for the plasma to reach the required temperature of the order of 100 million degrees and the ions must be confined long enough for a considerable fraction to fuse, so it is obvious that the larger τ_E, the more effective a fusion reactor will be as a source of power. The ideal situation is that no extra power is needed to supply heat to the plasma, in which case τ_E will become infinite and the plasma will burn until all fuel has been consumed.

The general idea is that outside energy is at first used to bring the plasma to the required temperature, and once the fusion reactions get going the α-particles provide more and more of the energy needed to keep the plasma at the right temperature. The Lawson criterion or, the virtually identical, ignition condition states that the "fusion triple product" $nT\,\tau_E$ of density, energy confinement time and plasma temperature must fulfil the condition:

$$nT\,\tau_E > 3 \times 10^{21}\,\text{m}^{-3}\text{keV\,s},$$

in units of *particles per cubic metre × kiloelectronvolts × seconds*. This is the most important condition in fusion. As an illustration to show how difficult it is to satisfy the Lawson criterion we note that the biggest fusion machine that has so far been operating, the Joint European Torus (JET) at Culham in the UK, achieved around one fifth of this value in its deuterium-tritium experiments in 1997 (see Chap. 9). It should also be noted that the Lawson condition by itself is not sufficient; each of the three quantities entering into the triple product must also be large enough. It is for instance not sufficient to heat the plasma to 1 billion degrees (far higher than needed for D-T fusion), with confinement time and density high enough to satisfy the above condition, but still too low to get fusion going.

The best route to ignition for a deuterium-tritium plasma turns out to be at around 10–20 keV (100–200 million degrees). The Lawson condition then

---

[1]Energy is expressed in joule, and power in joule per second, so this ratio has the dimension of time.

gives (with a confinement time of 5 s and a temperature of 10 keV) that the density of the plasma should be at least about 10^{20} particles per cubic metre. This is rather low: a factor of 100,000 less than the density of air, the number of particles in a cubic metre of dry air being about 2.5×10^{25} particles/m³. If the density were much higher, the outward pressure would also be higher, and it would be much more difficult to confine the plasma. On the other hand, if the density is too low, the rate at which fusion reactions would take place and energy can be generated will be too small to be of practical interest.

The fusion reactions will also cause the plasma composition to change appreciably as the products of the reaction (D + T → ⁴He (3.5 MeV) + n (14.1 MeV)) spread through the plasma. The neutrons, being neutral in charge, but highly energetic, are not affected by the magnetic fields and fly off. They must be absorbed somewhere, and their energy used for generating electricity, the eventual aim of any power plant. The much less energetic α-particles (⁴He) carry about 20% of the surplus energy of the reaction. Since they are positively charged, they are trapped by the magnetic field and stay in the plasma. They can be used to further heat the plasma or keep it at the right temperature. But they are also spent fuel, 'waste' products of the fusion processes, and should not become a major fraction of the plasma as it would reduce the fusion reaction rate. This is also happening in the Sun where hydrogen nuclei fuse (via the three-step process explained in Chap. 2) into helium. The helium 'ashes' are left behind and eventually the core becomes so depleted of fuel that the energy production will start to fall.

So, let's summarise what is minimally needed to get controlled fusion going and possibly turn it into a useful energy source:

1. a plasma at a temperature of a couple of hundred million degrees;
2. with a density that is not too high (about 1/10,000 of atmospheric pressure) to withstand thermodynamic pressures at such temperatures and to keep the fusion process under control;
3. a plasma that is very pure (i.e., consists only of deuterium and tritium), since radiation losses greatly increase with impurities (especially metallic impurities from the vacuum vessel wall, which have high atomic number Z and radiate much more than deuterium and tritium, which have low Z);
4. and that is confined long enough by a magnetic field (the magnetic pressure to compensate the thermodynamic pressure) for a considerable fraction of the deuterium-tritium ions to fuse and net energy gain to become possible; so that

5. after ignition, i.e. after the plasma starts to burn, more fusion energy is produced per unit time than the total of all losses (radiation, particle loss) and the process will sustain itself;
6. which still leaves the problem of extracting any surplus fusion energy from the reactor vessel and turning it into electricity.

Progress towards ignition is being made, but so far this goal has not been reached. Seventy years after Lawson formulated the condition discussed here and in spite of frequent promises by scientists and leaders of experiments, no fusion experiment has reached a situation in which the α-particle heating balances the energy losses, so that the fusion reaction can become self-sustaining.

Plasma Heating

How can a plasma be heated to the temperature needed for fusion reactions to start happening? We already know that a plasma can be formed by sending a current through a gas. The current will heat up the gas and strip the atoms of their electrons. This type of heating is called *ohmic* heating or resistance heating, the process by which an electric current through a conductor produces heat due to the resistance of the matter the conductor is made of. A copper wire for instance becomes hot when a current is passed through it, because the electrons in the wire collide with the much heavier ions and transfer heat to them. As we know from secondary school, the amount of heat is proportional to the product of the wire's resistance R and the square of the current I, i.e. proportional to $I^2 R$. In a plasma the number of collisions between ions and electrons is much smaller than in a wire, up to 10 orders of magnitude (10 billion times) smaller. On the other hand, the currents that can be driven through a plasma can be very large, more than 100,000 amperes, even many millions of amperes.

The first plasma was created by this method in 1947 at Imperial College in London in a precursor of the magnetic confinement machines in use today, a doughnut-shaped gas-filled glass vacuum vessel through which a current was sent. At the time it was not possible to measure the temperature of the plasma, and only a bright flash of light was seen in the gas, showing that the plasma was highly unstable and lasted only for a fraction of a second. As the plasma gets hotter and hotter, its electric resistance decreases (as noted before, the conductivity of a plasma is one of its most notable features; a plasma is almost a superconductor) and this type of heating becomes less effective; I^2R will go down. The maximum temperature that can be reached with ohmic heating is less than 50 million degrees, extremely hot by almost any standards, but insufficient for fusion. Ohmic heating is also limited by the fact that increasing the plasma current beyond a certain limit tends to disrupt the plasma.

To improve on this, two main heating techniques have been developed that go beyond ohmic heating:

(1) The first uses powerful beams of highly energetic neutral particles (deuterium atoms or a mixture of deuterium and tritium atoms) that are injected into the plasma. This is called neutral-beam heating. It involves a rather elaborate process whereby the deuterium atoms are first ionized (so that they become positively charged) and then accelerated to high energy by passing them through an electric field. The resulting beam of energetic charged particles cannot be directly injected into the plasma as the particles would be deflected by the magnetic field and harmful plasma instabilities would arise. So, they are first (re)neutralised by passing them through a deuterium gas from which they pick up electrons to recombine into neutral atoms. Then they are injected and, since they are now neutral, but still very energetic, can penetrate deep into the plasma, collide with many plasma particles and give up some of their energy, further heating the plasma. In the process they are again ionized and become part of the plasma. For this method to be useful, the injected particles must have very high velocities (more than 3,000 km/s) in order to penetrate into the centre of the plasma. If they are too slow, they will give up all or most of their energy before getting to the centre.

(2) The second approach works rather like the heating of food in a microwave oven. Electromagnetic waves (typically radio waves) are launched into the plasma at different frequencies. The trick is to tune the frequencies of these waves to some natural resonance frequency of the ions and electrons in the magnetic field holding the plasma together, such as the orbiting frequencies of the particles in the field. Since ions and electrons have vastly different masses, the resonance frequencies are also vastly different, at about 50 MHz and 170 GHz, respectively. At such frequencies, the energy carried by the waves is transferred to the charged particles, increasing their velocity, and at the same time their temperature and the temperature of the plasma as a whole. The radio waves, typically of wavelengths of several metres in free space (for ions) and a few mm (for electrons), are launched from an antenna or waveguide at the edge of the plasma. Such heating is called ion cyclotron resonance heating

(ICRH), lower hybrid (LH) (resonance) heating, or electron cyclotron resonance heating (ECRH).

The use of these techniques was no plain sailing. The temperature rise was less than expected, as energy losses also increased when increasing the heating power. It turned out that the situation was like "a house with central heating where the windows open more and more as the heating is turned up. The room can still be heated, but it takes more energy than if the windows stay closed." As we will see later in this book, a machine like ITER will use a combination of these heating techniques: neutral-beam injection and two sources of high-frequency electromagnetic waves, with antennae the size of a bus.

Plasma Instabilities and Disruptions

Energy is stored in the plasma, both in the confining magnetic field and in the plasma particles. Under certain conditions instabilities can arise and grow rapidly by extracting energy stored in the plasma. Instabilities and the ensuing disruptions can cause the plasma to crash into the wall of the containing vessel. The resulting catastrophic loss of plasma combined with the potential damage to the wall has led to the realization that in a fusion reactor such instabilities must be avoided, and that stability is crucial. But how to do this as, due to the turbulent behaviour of the hot fusion plasma, instabilities seem to be part and parcel of it and plasma containment is greatly hampered by hundreds of different instabilities?

Disruptions, still poorly understood, are characterised by very rapid growth, and give rise to rather violent effects, causing the current to abruptly terminate, which results in the loss of temperature and confinement. Because of the short time in which the thermal and electromagnetic energy of the plasma is released, large forces act on the plasma-containing vacuum vessel, putting it under huge mechanical strain. All the energy is dumped into the structure of the machine, and in big machines forces of several hundred tons have been measured. Such events have been witnessed both at the JET tokamak in Britain, where it caused the whole giant machine to jump a few centimetres into the air, and in the *Tokamak Fusion Test Reactor* at Princeton (USA), producing "a noise that sounded like the hammer in hell, followed by echoing thunder." It is not uncommon for disruptions to cause considerable damage, and prevention or mitigation is essential. An eventual power plant will have to be able to withstand forces of at least an order of magnitude higher than estimated to be unleashed by a possible disruption to make sure that the construction is safe, which adds considerably to the construction costs.

Disruptions occur when the plasma is pushed too close to the limits for a plasma in the device, such as density and pressure. How this happens is not entirely clear, so the approach to tackle them is avoidance (by staying away from the plasma limits), prediction (by employing sensors) and amelioration (by automatic controls that change the plasma parameters). The problem is a potential showstopper for fusion and is receiving a great deal of attention.

Disruptions and plasma instabilities are very closely related, and the latter are often the cause of disruptions. Instabilities have been plaguing fusion research since its early beginning, and new and more vicious instabilities keep popping up. It will be a tremendous challenge for fusion research to find sufficiently stable plasma configurations.

All instabilities are due to small but exponentially growing imperfections or noise in the system and are therefore unavoidable since such imperfections are always present. The huge range of instabilities is well illustrated by the list of more than 50 plasma instabilities, from Buneman instability to Weibel instability, in *Wikipedia*. The list is incomplete as one of the most recent and most troublesome kinds of instabilities, the edge-localised modes (ELMs), is not even mentioned. There are many types of ELMs, but the most common one is closely related to ballooning instabilities which act like the elongations formed in a long balloon when it is squeezed.

To build proper theories of such instabilities nonlinear analysis is needed, and as can be imagined, this poses difficult, almost insurmountable problems. All physical processes are nonlinear to a certain extent, but normally the nonlinearity is small. One can then start with a linear analysis and try to calculate nonlinear effects as small perturbations. Such an approach is no longer valid for the turbulent behaviour in a plasma, which is nonlinear in the extreme, a situation in which fluctuations can no longer be treated as small perturbations.

In general, plasma stability is improved by limiting the pressure or current through the device. Stability theory is thus concerned with two basic problems: how can the actual limits on pressure and current be calculated for any given magnetic configuration, and how can the magnetic configuration be optimised so that the pressure and current limits are as high as possible?

In the situation of a well-confined plasma in equilibrium, separated from the wall of the containment vessel by a vacuum region, instabilities can be classified into internal and external instabilities, based on whether or not the surface of the plasma moves as the instability grows. An internal instability implies that the plasma surface remains in place and the instability occurs purely within the plasma. Often, they do not lead to catastrophic loss of plasma, as the plasma remains contained. External instabilities, on the other hand, involve motion of the plasma surface, and hence the entire plasma. Such motion may lead to the plasma striking the wall, ending the plasma's existence. In a fusion plasma external modes are particularly dangerous.

A second way to classify plasma instabilities is by the source driving the instability. In general, both perpendicular and parallel currents (relative to the magnetic field) occur in a plasma and each of these can drive instabilities. Instabilities driven by perpendicular currents are often called "pressure-driven" modes. Pressure-driven instabilities are usually internal modes and set an important limit on some plasma parameters, such as the thermal plasma pressure or the magnetic pressure.

Instabilities driven by parallel currents are often called "current-driven" modes. A common name for current-driven instabilities is also "*kink modes*," because the plasma deforms into a kink-like shape. Kink modes can be either internal or external.

In certain situations, the parallel and perpendicular currents combine to drive an instability, also referred to as the "*ballooning-kink mode*". This is usually the most dangerous mode in a fusion plasma. It is an external mode, implying that it can lead to a rapid loss of plasma energy and plasma current to the wall of the containment vessel. Another name for it is "*explosive instability*", the narrow fingers of plasma (ballooning fingers) produced by the instability are capable of accelerating and pushing aside the surrounding magnetic field, causing a sudden, explosive release of energy. Other ballooning instabilities go under the names of *peeling-ballooning* and the already mentioned ELMs. There are many types of the latter modes, which are local instabilities that take the form of filamentary eruptions, ballooning out from the plasma towards the wall of the reactor vessel. It is a way for the plasma to let off steam when the pressure in the plasma is building up.

In the first fusion devices, constructed in the late 1940s to early 1950s, it was soon found that the plasma becomes dreadfully unstable when attempts are made to extend the plasma confinement time beyond a few microseconds. Theoretical analysis showed that several types of pressure-driven and current driven instabilities occurred, going by the names (although the use of these terms does not seem to be consistent) of kink instability, representing transverse displacements of the beam cross-section, changing only the position of the beam's centre of mass (Fig. 5.3a), sausage instability, resulting from

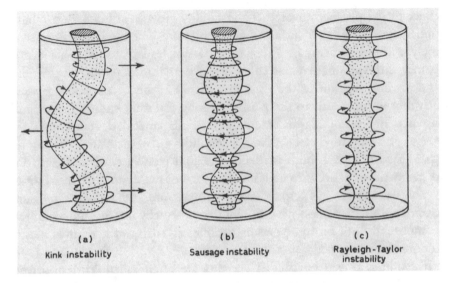

(a) Kink instability

(b) Sausage instability

(c) Rayleigh-Taylor instability

Fig. 5.3 Schematic representation of a few simple instabilities

variations of the beam radius with distance along the beam axis (the plasma was pinched unevenly at one or several places along its length and broke up into a series of blobs like a string of sausages) (Fig. 5.3b), and Rayleigh-Taylor instability (Fig. 5.3c).

The latter is a classical instability first observed in fluid dynamics. It concerns the instability of an interface between two media, normally fluids of different densities, which occurs when the light fluid is pushing the heavy fluid. To explain this instability, we can look at two fluids that do not mix, for instance water and oil. Water is denser than oil, so in a normal situation oil will float on water and not the other way round. The situation where the water is on top can briefly be in equilibrium, but the equilibrium is unstable, and any perturbation of the water-oil interface will cause the situation to be reversed. If for some reason a drop of water is displaced downwards with an equal volume of oil displaced upwards, the disturbance will grow extremely fast, moving the water down and displacing the oil upwards, thus restoring the normal stable configuration. The instability as described here is driven by gravity.

In a plasma the instability cannot be driven by gravity as the plasma weighs almost nothing and gravitational forces are rarely of much importance in plasmas. Here, pressure on the plasma-vacuum interface is the culprit. The vacuum only contains the magnetic field, while in the plasma (consisting of charged particles), there are both electric and magnetic fields. The magnetic field exerts a magnetic pressure (due to the energy density of the magnetic field) on the plasma, and since the magnetic field is curved (in most doughnut or torus shaped devices) this pressure is not uniform and under its action the plasma and the magnetic field try to exchange position. They are the equivalents of the fluids with different densities in the ordinary Rayleigh-Taylor instability.

6

Early History and Declassification

The basic questions that the early pioneers in fusion research had to answer were how to make a plasma and keep it in place for some (very short) time, and what kind of device could achieve this. It became soon clear that, if anything could do the job, it would be magnetic fields, but in what kind of vessel?

The simplest arrangement would be to take a straight tube. A magnetic field can be induced by a current running through wires wound around the tube. Such a field would have the major advantage that it is uniform, i.e., it would have the same strength everywhere along the tube, greatly improving the chances of stability. It is also clear though that such a tube, open at both ends, is unsuitable as the particles would eventually escape at the ends and fly off. Nonetheless, people have managed to design devices that to some extent manage to plug the endpoints of such a tube. These are called mirror devices and will be discussed briefly below, but in the designs that were most successful the tube is bent into a ring or doughnut, sacrificing the uniformity of the magnetic field. The most successful device of this type is the Soviet-invented tokamak, which will receive ample coverage in the rest of the book. Before going into more detail, we will first look at the general setup of such devices.

© The Author(s), under exclusive license to Springer Nature
Switzerland AG 2021
L. J. Reinders, *Sun in a Bottle?... Pie in the Sky!*,
https://doi.org/10.1007/978-3-030-74734-3_6

General Properties of Toroidal Devices

The proper geometric term for a linear tube bent into a circle is a torus. It is just a tube in the form of a hollow ring. All currently used magnetic fusion devices have such a torus shape. Figure 6.1 shows a schematic representation, with the major axis through the centre of the torus, called the z-axis, as indicated. The figure also shows the major radius R (the distance from the centre of the torus to the centre of the tube) and the minor radius a (the radius of the tube). The ratio of these two radii, R/a, is an important quantity and is called the aspect ratio of the torus.

A magnetic field generated by coils wound around the torus (toroidal field) will create lines of force (magnetic field lines) going round in the torus, and charged particles will travel in circles around the z-axis (while describing their corkscrew trajectories along the field lines in the torus due to the Lorentz force, see Chap. 5). However, the deviation from a straight cylinder caused by bending the cylinder into a ring implies that the magnetic field is no longer truly uniform. It is stronger on the inside of the torus where the magnetic field lines are closer together than on the outside. This results in particles drifting upwards and downwards depending on their charge (a problem common to all torus-like devices). This drift is called the toroidal drift. The resulting separation of charges will induce an electric field causing the particles to drift further and further apart and eventually hit the wall of the torus. Therefore this setup cannot work unless the charge separation is cancelled.

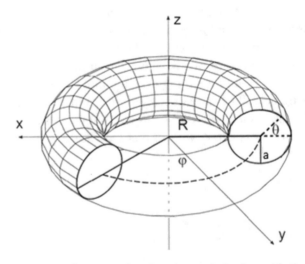

Fig. 6.1 Schematic view of a torus, showing the z-axis (major axis), the major radius R and the minor radius a

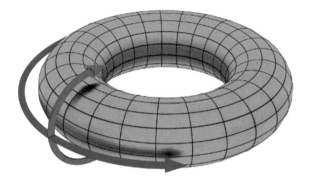

Fig. 6.2 Illustration of toroidal (blue arrow) and poloidal (red arrow). *Wikipedia*

The word 'toroidal', which often appears in discussions of torus-shaped confinement systems, indicates that the magnetic field or current goes the long way round in the torus, as indicated by the blue arrow in Fig. 6.2. Something that goes round the short way is called poloidal, as indicated by the red arrow in the figure.

For some designs a solution to cancel the toroidal drift was found by sending a current through the torus, which produces a poloidal magnetic field with its field lines going the short away round the torus. (The current itself is a toroidal current as it goes the long way round the torus.) This poloidal field, produced by the current, combines with the toroidal field from the coils on the outside of the torus. The resulting combined field will be twisted into a helix, so that any given particle travelling around in the torus will find itself alternately on the outside and the inside of the torus, and the drift will be cancelled out. Figure 6.3 shows graphically how this twisted field comes about.

Let us explore in a little more detail how a plasma particle can be confined by a magnetic field in such a toroidal device. Consider a cross-section at some point through the tube shown in Fig. 6.1 or 6.3, perpendicular to the magnetic field. Assuming that the magnetic field is nonzero at all points in this circular cross-sectional plane, a magnetic field line will pass through each point of the plane. Each time a magnetic field line has gone around the tube, the point at which it will go through the plane will be slightly displaced, slowly rotating about the axis of the tube and forming an entire closed surface, called a "magnetic surface". Ideally the magnetic field lines will describe closed paths. In this configuration this is only true for the field line going through the axis, the centre of this cross-sectional plane. All other field lines will rotate around this axis and when the parameters are chosen properly will form closed magnetic surfaces that remain completely within the tube.

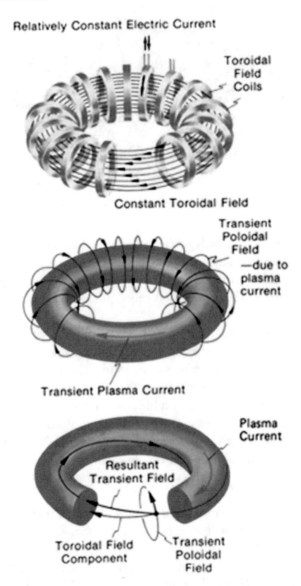

Fig. 6.3 Magnetic fields and current in a torus. In blue in the top figure are the toroidal field coils which produce a field going the long way round the torus (toroidal field). The middle figure shows the plasma current (red) and the resulting poloidal field. The bottom figure combines the two fields into the resultant helical (twisted) field, cancelling out any drift. *Wikipedia*

The particles move on such surfaces, preventing them from crashing into the wall. For the plasma to remain confined inside the tube and not to crash into the wall, the tube must enclose an entire family of such magnetic surfaces.

Pinches, Stellarators and Mirrors

In the first few years after World War II, nuclear research in the US and the Soviet Union remained preoccupied with building bombs, and nuclear fusion for energy-generating purposes did not get much attention. Consequently, the first research on fusion was carried out in the late 1940s in the UK, first at Imperial College in London and at Oxford University, later at the Atomic Energy Research Establishment (AERE), founded in 1946 at Harwell, just 25 kms from Oxford.

The earliest proposals made for fusion devices are so-called **pinch devices**. When a current of sufficient strength is passed through a gas, it ionizes the gas, raises the temperature (ohmic heating) and produces a circular (poloidal) magnetic field surrounding the current. This is a wonderfully simple idea: the plasma current, which heats the plasma, produces its own magnetic field to confine the plasma. If the current is raised further, the degree of ionization and the temperature increase, but also the strength of the magnetic field. This field exerts a force on the column of ionized gas (the plasma) and compresses it, i.e., "pinches" it into a thin filament. The idea is that this will pull the material together, so that the nuclei can get close enough to fuse.

In principle this should work for a straight tube, but to prevent the plasma from escaping at the endpoints the tube has to be bent into a torus. To prevent the toroidal drift, as discussed above, a vertical magnetic field was applied that would interact with the current flowing in the tube.

Encouraging results with such toroidal devices were soon achieved: pinched plasma that lasted for one ten-thousandth of a second. When scaling up to larger tori, creating ever greater plasma currents, it was soon found that the plasma was dreadfully unstable and started to wriggle about like a snake until it made contact with the vessel wall and was extinguished. The kink instability had made its appearance, showing its wavy behaviour: crosswise displacements of the beam without changing the form of the beam other than the position of its centre of mass (Fig. 5.3a). Such instabilities are a natural consequence of the pinch effect. The magnetic field lines induced by the current can be seen as a series of evenly spaced rings around the current. Any slight kink in the current will disturb this equal spacing, causing the field lines to be more closely packed together on one side and more widely apart on the other. More closely packed field lines mean a stronger magnetic field, disturbing the balance of forces and reinforcing the kink, causing the whole thing to get out of control and the pinched plasma to crash into the wall of the containing vessel.

In spite of a lack of understanding of such instabilities, there was an ever increasing pressure at the AERE to build larger machines that could produce the temperatures necessary for fusion. This resulted in the design of the ZETA device, Zero Energy Thermonuclear Assembly, an apt name as in its first runs it produced just 10^{-12} (one million millionth) of the energy that it consumed. It was the biggest fusion experiment so far. The problem with the kink instability was solved by applying another magnetic field going around the torus and exerting a force that would push the kink back into line. In August 1957 ZETA was fired up for the first time and soon started to register millions of neutrons at a temperature of at least 1 million degrees, and probably 5 million. The press got wind of this and soon the newspapers were full of wild, overblown stories about Britain's fusion success. However, the triumph proved to be brief. The neutrons had been wrongly interpreted as originating from nuclear fusion events. If they originated from thermonuclear reactions, they would be flying out equally in all directions and with similar energies. The ZETA team had postponed measuring the neutrons' directions on the ground that there were too few of them. Others soon made the measurements and showed that the majority of the neutrons were of spurious origin. The "mighty ZETA", as the Daily Mail had called it, had crashed. The neutrons were not from fusion reactions, but a by-product of plasma heating. It turned out that both the Russians and the Americans had also observed such spurious neutrons a few years earlier but had been more careful and had not let themselves be carried away as the British had.

Experiments on the ZETA continued until 1968 when the machine was shut down. In the meantime, in 1965, the British had opened a dedicated nuclear fusion laboratory in Culham, nearby in Oxfordshire. In the first two decades of its existence Culham built almost 30 different experiments trying out a variety of fusion concepts.

In the US, fusion energy had been discussed from time to time within the framework of building the hydrogen bomb, but then in the very early 1950s, three separate groups, each pursuing a different route to fusion, sprang up almost simultaneously at Los Alamos, Princeton University, and the Livermore branch of the University of California's Radiation Laboratory.

One striking feature of the programme's first years was the astonishingly short period considered necessary for establishing the feasibility of fusion as an energy source. It was estimated that 3–4 years and 1 million dollars would suffice to find out if a hot plasma could be contained. Today it is clear that this estimate was vastly wrong, so wrong that one wonders how such a huge mistake could have been made.

In 1952, on the matter of fusion, there was room for wide extremes of optimism and scepticism. Almost nothing was known about the behaviour of plasma at the extreme temperatures required for fusion. A scientific basis for fusion hardly existed, while the situation with respect to technology was even more wide open. Magnetic fields of the strength needed had never before been created, nor had plasmas with temperatures of millions of degrees ever been made in the laboratory. No one really knew how to make them. The fusion scientists all took the stance of optimism, but just for their own particular design and on what basis they did so was not always clear. They were all seeking public funds to chase after a wild, poorly thought-out idea. Social factors probably also played a role. Post-war exuberance had created an atmosphere in which everything seemed possible, that anything could be done if only the decision was taken to make it a matter of national priority. Overconfidence, perhaps, after the successful completion of the Manhattan project, which several of the fusion pioneers had participated in or been in contact with. Another factor is that there was no one who knew all the bits and pieces of scientific and engineering knowhow underlying fusion. No one had an overview. There were no established disciplines in this field. There was no fusion community yet. There were no nuclear engineers. The participants came from various fields of physics. But one would also have thought that this lack of established practice would have made them more cautious in making predictions and claims in uncharted territory.

Not everyone expressed the same optimism though. Edwin McMillan (1907–1991), an American physicist and Nobel laureate, for instance, did not work on fusion, but pointed out in 1952 that "*to confine a plasma with a magnetic field was to place it in a distinctly unnatural situation, somewhat like the situation that would result from trying to push all the water in one's bath to one side of the tub and keep it there.*" The plasma, like the water, would try every permissible mechanism to escape. And Edward Teller (1908–2003), in a now famous talk at a conference in Princeton in 1954, pointed out that any device with magnetic field lines that are convex towards the plasma would likely be unstable. All the devices considered by the American (and British) groups had such convex magnetic field lines and would suffer from these instabilities.

Pinch devices, similar to the ones used in Britain, were pursued at Los Alamos, but eventually found not to bring any effective results. It turned out not to be possible to keep the plasma stable, as the current passing through the plasma readily results in instabilities, while there were further problems with the plasma slowly leaking away. Stability remained the overriding problem, not only for the pinch devices but also for the other two approaches to be discussed below, and for that matter for present-day devices.

The second approach, pioneered in the US, was the so-called **stellarator** (a name chosen to refer to stellar processes) invented by Lyman Spitzer (1914–1997), a professor of astronomy at Princeton University. Spitzer quickly realised that only a magnetic field could do the confinement trick but did not believe that the pinch effect was the answer to magnetic confinement. A pinch device, if it were to work, would deliver neutrons from fusion reactions in fast pulses, providing energy in rapidly succeeding bursts, while Spitzer rightly thought that only a steady-state device, which would be able to provide energy continuously, would be practical. He discarded the idea of sending a current through the plasma to generate a confining magnetic field, but envisaged a single coil system to produce a magnetic field to confine the plasma.

A uniform magnetic field can be created in a tube by winding a wire round the tube along its full length and sending a current through the wire. Since a tube of finite length is no good, Spitzer bent it into a doughnut shape (torus) and, as in pinch devices, was confronted with particle drift, resulting in a separation of positively and negatively charged particles. This separation gives rise to an electric field that neutralises the confining effect of the magnetic field and sweeps the ionised gas towards the wall of the tube before having even once circled the torus. So instead of a torus he built a figure-8 shaped device consisting of two crossing straight sections joined by curved ends (Fig. 6.4), essentially a double torus, with magnetic fields in opposite directions, linked to form a kind of pretzel. With this modification, a particle that tends to drift upwards at one curved end will tend to drift downwards at the other end, cancelling out the net drift.

As explained above, the magnetic field lines go around in the tube and slowly rotate around the axis, with the angle of rotation determined by the

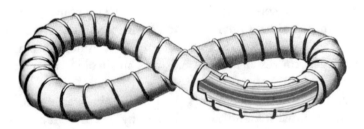

Fig. 6.4 Simplest stellarator design

twist of the figure-8 configuration. The tube can be designed to give any desired rotational angle. The essential idea of the stellarator is that the helical twisting of the magnetic field lines is achieved solely with external coils; no current is sent through the plasma.

Spitzer started with a table-top device to show that the plasma could be created and confined and the electrons in the plasma heated to 1 million degrees, to be followed by a larger Model B which would also heat the plasma ions to 1 million degrees. The final Model C would then be a virtual prototype power reactor, reaching thermonuclear temperatures of 100 million degrees. He also thought that the whole project would take about a decade. An astonishingly optimistic attitude, certainly since he was aware that there could be many difficulties ahead, in the form of instabilities and other unwanted behaviour. In one of his papers he wrote: "*Little is known about the instability of plasma oscillations in otherwise quiescent plasma*" and "*In a system with so many particles and with so few collisions as a stellarator plasma, almost any type of behaviour would not be too surprising.*"

The Model-A stellarator was ready for work in the course of 1952. Its aim was to create a fully ionised plasma and heat its electrons to 1 million degrees. The device was too small to heat the rest of the plasma to this temperature, but it proved that the figure-8 geometry could indeed make plasmas more easily (at lower voltage and magnetic field) than was possible in a simple torus. Model B, which was not much larger than Model A, was designed for a stronger magnetic field, by a factor of about 50. It was hoped that it would confine the ions long enough (about 30 ms) to heat them too to one million degrees.

Difficulties with the original version of the model and new developments resulted in the construction of a series of improved versions. The models all suffered from loss of plasma at rates far worse than theoretical predictions. This situation, i.e., theoretical predictions not being borne out by experiment, turns out be a common theme in fusion research.

Planning for Model C began already in 1954, even before Model B had produced any results, and consequently was far too optimistic. It was designed to "*serve partly as a research facility, partly as a prototype or pilot plant for a full-scale power producing reactor.*" And, still more optimistically, such a full-scale reactor was also already on the drawing boards, called Model D, its design being ready at the end of 1954: a 165 m long machine in the form of a large figure-8 operating at a temperature of about 200 million degrees. It would require an investment of close to 1 billion dollars (1950s dollars!) and was designed to be capable of producing a net electrical power output of 5 gigawatt, enough for 5 million households and about 10 times the output of large power plants in operation at that time. All this happened before Teller gave his famous talk predicting instabilities in devices like the stellarator. Work on the smaller models and on other toroidal systems established that several of the physics and technology assumptions of the Model-D design were untenable. The work on Model D was stopped before it had even started. Model D has never risen from its slumber, although its design set a precedent for future stellarator design.

By the time Model C started operation in 1961, it was becoming clear that it would not be able to produce any large-scale fusion. A principal finding over a broad range of experiments on Model C was that confinement was linearly proportional to the applied magnetic field strength, instead of having the expected quadratic dependence. A linear dependence would require too high magnetic fields to make confinement practical. Model C continued in operation until the facility went under in the tokamak stampede in 1969 and stellarator research at Princeton was abandoned.

The third approach consisted of magnetic **mirrors**. These are open-ended linear confinement systems, in principle the most ideal geometry for plasma confinement one can think of (in essence a straight tube). The basic design is shown in Fig. 6.5. It is essentially a bunch of magnetic field lines, bundled together at each end of the tube, hence the magnetic density or field strength is higher at these end points than in the middle. When plasma moves from lower to higher density fields, it can reflect. This reflection is known as a plasma mirror (a phenomenon that has an astrophysical equivalent). The mirror device was envisaged to have two plasma mirrors facing each other and the goal was to have plasma ricochet back and forth until it slammed together in the centre and fused.

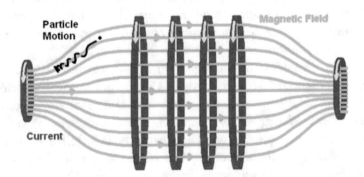

Fig. 6.5 Basic magnetic mirror. The black wiggly line shows the motion of a charged particle along a magnetic field line. The (optional) rings in the centre extend the confinement area horizontally

A charged particle will travel in a helical motion around a magnetic field line, as shown by the black wiggly line in Fig. 6.5, and will try to flow out of the system at the end. When approaching the end, it will encounter an increasingly strong magnetic field and the radius of its motion will become ever smaller. It will start to gyrate faster and faster, increasing its rotational energy. Since energy is conserved, the particle has to take the extra rotational energy out of its translational motion and will consequently slow down its efforts to escape from the end. This will eventually cause it to reverse direction (as if reflecting from a mirror) and return to the confinement area.

So, the idea of mirror confinement is to confine the particles in a magnetic field in a kind of bottle by making the field stronger at the ends. This system can easily be modelled, the boundary conditions written down, and the equations solved. The math will tell you that everything works out fine. Except it doesn't. In real life, mirrors have problems. The plasma leaks out at the ends and it is very hard to ensure that no plasma is leaking away. On paper, everything looks fine, but in reality that is not so. This highlights an important deficiency of plasma modelling. The simpler the math, the more it deviates from real life.

The mirror device is essentially a leaking magnetic bottle. This problem would eventually be the death knell for these systems, and we will not dwell on these developments in further detail. In spite of this drawback, studies on magnetic mirrors were conducted in a large number of laboratories and in the early days were at least as important in overall fusion research as the work on closed systems. Various ingenious methods were devised to plug the endpoints to try to stop particles leaking away. It culminated in the construction of the Mirror Fusion Test Facility (MFTF) at a cost of 372 million dollars at the Livermore laboratory, at that time the most expensive project in its history. It opened early in 1986 and was promptly shut down. The reason given was to balance the US federal budget, but it had already become clear that as a power-producing machine the mirror appeared to be a dead end and would never be able to beat the tokamak. There was no point in continuing with such devices, which soon meant the end of mirror research in the US.

In Europe, excluding Britain, nuclear research was coordinated almost from the start by the European Atomic Energy Community (Euratom), established in 1957 by the six founding members of what is now the European Union (EU). Britain only joined Euratom after its accession to the European Community in 1973. Fusion seemed ideal as a field of European cooperation and one advantage in this respect was that there were no questions related to industrial and military applications (as in the case of fission).

All fusion research in Europe evolved through association agreements with Euratom, and Euratom would finance, develop, coordinate and supervise national fusion programmes. These agreements were a great success, because of their simplicity and effectiveness. A laboratory that accepted the Community rules on tendering, knowledge sharing and work evaluation could have 25% of its general expenditure paid by Euratom, without losing any of its autonomy. This was a very advantageous arrangement and most countries quickly signed up for it.

While in 1958 experimental research on a large scale was conducted mostly in Britain as far as Europe is concerned, in 1965 France, Germany and some other countries, all working under association treaties with Euratom, had entered the fray and were actively involved in such research.

Germany's status as an occupied territory after World War II formally ended when the General Treaty, signed in 1952 by West Germany and the Western Allies (Britain, France and the US), came into force in 1955. Once restrictions on applied nuclear physics work were lifted, this kind of research really took off in Germany and within a decade it was the world's leading exporter of nuclear technology.

Because of the secrecy adhered to in the field, Germany had not been aware of the extent and orientation of fusion research in the US, UK and Soviet Union. The declassification (see below) coinciding with the 2nd Geneva Conference on the Peaceful Uses of Atomic Energy in 1958 had a catalysing impact in this respect. The German effort really took off in 1960 with the foundation of the Max Planck Institute for Plasma Physics (IPP) in Garching, near Munich. A few years later the institute was integrated into the European Fusion Program through an association agreement with Euratom. With a staff of more than 1000 people the IPP is currently one of the biggest fusion research institutes in Europe. The objective of the institute is to carry out research in plasma physics and adjacent fields, as well as to develop the methods and tools needed for such research. As in its name, no allusion was made to the eventual goal of building a fusion reactor; even fusion research was not mentioned. Throughout its early history IPP took the developments at Princeton as its guideline, and so the institute's foundation year already saw the launch of its first stellarator, which would eventually culminate in the Wendelstein 7-X, to be discussed in a later chapter.

France's nuclear fusion programme started with the formation of a working group for theoretical research after Bhabha's prediction at the 1955 Geneva meeting (see Chap. 1). Proper research into fusion began in 1957 with a first torus-like device at Fontenay-aux-Roses. In the first years, mirror devices were also the focus of French efforts.

Other European countries that carried out fusion experiments from quite early in the game were Italy, the Netherlands, Denmark and Sweden.

Declassification

After World War II America greatly tightened security around all nuclear efforts, irrespective of whether they had to do with fission or fusion, excluding all foreigners, even former allies like Britain and Canada, from access to its nuclear research.

The first calls for declassification in the US date from 1952 when people were arguing that secrecy hampered the recruitment of capable people, the free flow of information and ideas, and the publication of papers that could be read, discussed and criticized by others working outside or inside the field.

Although most scientists were in general strongly in favour of declassification, the Atomic Energy Commission (AEC) decided against it. The mood for declassification also fluctuated with the perceived state of affairs. The early period of fusion research from 1952 to 1958 was characterised by wild swings from optimism to pessimism about imminent success and back. Fusion and plasma physics were not (yet) a normal scientific discipline and at least in the US it worked on a trial-and-error invention cycle, often based on back-of-the-envelope calculations. Scientists and administrators were convinced that the nuclear fusion nut could be cracked in a matter of a few years. In such a situation declassification would of course not be opportune as it might enable others and, God forbid, maybe even foreigners, to reap (part of) the glory.

Declassification also became a topic in the discussion on the peaceful uses of atomic energy, either from fission or from fusion. After all, apart from producing weapons nuclear power can also be used to generate electricity and all nations in the world were entitled to benefit from such peaceful use. The discussion was started by President Eisenhower in his Atoms for Peace address to the United Nations General Assembly in December 1953. He called for a conference on such peaceful uses to be held in Geneva and proposed that an international agency be established for making information and nuclear material available under strict guidelines to other nations wishing to engage in peaceful applications of nuclear technology.

The Russians initially rejected Eisenhower's proposals as empty propaganda, but a few months later the Soviet Union showed some willingness to start negotiations with the Americans and to participate in some international organisation. It resulted in 1955 in the first United Nations International Conference on the Peaceful Uses of Atomic Energy and in 1957 in the creation of the International Atomic Energy Agency (IAEA). The main topics of the conference were, for that matter, not nuclear fusion and/or nuclear fusion reactors, as it was much too early for that, but power generation in nuclear fission reactors.

The lifting of secrecy of nuclear fusion research was finally decided on in April 1956 within the framework of the Canadian-British-American declassification conference, and in July 1956 the first review article on the subject appeared.

In the Soviet Union research in all nuclear fields was of course also secret, if possible even more secret. In the West not much was known about the Soviet efforts, while all kinds of rumours were circulating that the Russians had solved the problem of plasma heating and were well on their way to a working fusion reactor. Already in March 1954, just one year after Stalin's death and after the removal of Lavrentiy Beria, who so far had been the boss of everything nuclear in the Soviet Union, Soviet scientists had written to their government about the dangers of atmospheric nuclear testing because of radioactive fallout and urged the declassification of fusion research.

Declassification culminated in the 1958 2nd Geneva Conference, which like the first one was not a conference solely on nuclear fusion, but covered everything nuclear (physics, controlled fusion, fission, nuclear materials, fission reactors, applications, etc.). As far as nuclear fusion was concerned it became more an exercise in boasting and propaganda than a serious conference. The conference was open to the general public and included a giant exhibition where the various countries could vie with each other for the public's attention with their fancy exhibits. None of the teams working on fusion had essentially anything to show for, so it made sense to try to impress the public, which the Russians did by showing a full-scale model of one of their sputnik satellites, although it had of course nothing to do with fusion. As far as nuclear fusion was concerned, the Americans stole the limelight with a smartly polished exhibit of the Spitzer stellarator, for which no expense had been spared, but was nonetheless not working.

In spite of the suspicion, bravura and rivalry the conference was also a celebration of openness. Scientists could talk openly with each other and exchange ideas. They became aware that, in spite of the almost complete isolation the various parties had worked in, they had all invented circular and linear devices using magnetic fields and had all encountered the same difficulties and seemingly insurmountable problems, especially with instabilities.

The whole exercise showed how immature the field still was and the atmosphere of showmanship, exaggeration and hullabaloo present at the 1958 conference has remained one of the hallmarks of the nuclear fusion enterprise ever since.

7

Birth of the Tokamak

The tokamak is currently the most common torus-shaped device that uses magnetic fields to confine a plasma. It is the prime candidate for a future fusion reactor. The word is derived from the Russian acronym for toroidal chamber with magnetic coils (*toroidal'naja kamera s magnitnymi katushkami*).

The first design stems from work by Igor Tamm (1895–1971) and Andrei Sakharov (1921–1989) in the Soviet Union in the early 1950s. Already in October 1950, they reported the principal design of a magnetic thermonuclear reactor to a high Soviet official. This first report was followed by further communications in which the construction of a laboratory model was urged. In 1957 a more specific version was given the name tokamak.

Sakharov proposed to confine the plasma in a doughnut-shaped device (torus) by means of a toroidal magnetic field. He was also aware of the toroidal drift of the plasma particles due to the nonuniformity of the magnetic field and knew that a toroidal field alone, hence a field going around the torus the long way, cannot confine a plasma. As explained in Chap. 6, a solution was found by sending a current through the torus, which would produce a poloidal magnetic field, with its field lines going around the torus the short away. The poloidal and toroidal fields combined into a helically twisted field and cancelled the drift. This is essentially how a tokamak works, essentially the same as the pinch devices we already met, but with an important difference, as will be set out below.

From 1955 to 1969 the Soviet Union explored a whole range of devices (not only tokamaks) with the main objective of creating high temperatures.

© The Author(s), under exclusive license to Springer Nature Switzerland AG 2021
L. J. Reinders, *Sun in a Bottle?... Pie in the Sky!*,
https://doi.org/10.1007/978-3-030-74734-3_7

As in the UK and the US, 'false' neutrons were observed (already in 1952) and the kink/sausage instability threw a similar spanner in the works. Stabilization and other problems with early designs diverted the attention back to the original Sakharov-Tamm idea of using both a toroidal magnetic field and a toroidal current.

The function of the current was to provide equilibrium and plasma confinement, while the magnetic field was supposed to create stability, i.e., keep the kink instability in check as long as the current is not too high. It was found that the onset of the instability depends on the strength of the poloidal field relative to the main magnetic field, the toroidal field. The toroidal field has a stabilizing influence on the instability and can keep it in check if the current that induces the poloidal field is not too strong. This important insight is, as far as I know, due to the Russian theoretical physicist Vitaly Shafranov (1929–2014), and the ensuing limit to the plasma current is called the Kruskal-Shafranov limit. Let us consider this in a little more detail as it is a fundamental and trail-blazing discovery that has drastically changed the course of fusion research!

For tokamaks the ratio of the number of times a toroidal magnetic field line goes around in the torus the long way to the number of times a poloidal magnetic field line goes around the short way is defined as the **safety factor** q. The critical current corresponds to a combination of toroidal and poloidal fields whereby the plasma twisted by the magnetic fields comes back to exactly the same spot after having travelled once around the torus, i.e., a combination of fields for which the safety factor $q = 1$. In terms of magnetic field lines, $q = 1$ corresponds to the situation whereby a poloidal field line goes around the torus the short way exactly once for each time a toroidal field line goes around the long way. If the poloidal field line goes around more often it will twist the toroidal field too much, and hence the plasma too, and the situation becomes unstable ($q < 1$). If it goes around less often, no kink will develop ($q > 1$). So, q is a measure of the twist of the magnetic field and, at the same time, of the stability of the setup. By arranging the reactor such that this q is always greater than 1, tokamaks strongly suppress the instabilities that plagued other designs.

This led to the construction of the first proper tokamak in the world: the T-1 (Fig. 7.1), which started to operate in 1958. It was in this experiment at the Kurchatov Institute in Moscow that an abrupt change in stability was observed when the safety factor q became larger than 1. From 1955 onwards a whole series of facilities of this type were built. The last one in the series and in the Soviet Union was the T-15, which started to operate in 1988.

Fig. 7.1 Picture of the first tokamak in the world, the T-1, at the Kurchatov Institute

For all intents and purposes the physical layout of the tokamak is identical to a toroidal pinch device. The main difference with a toroidal pinch is the relative strengths of the magnetic fields. In a pinch device the poloidal field (responsible for the plasma current and the pinch) is much stronger than the toroidal field. In a pinch device the poloidal field must be strong in order achieve an appreciable pinch. In the tokamak it is just the other way round. The toroidal field in a tokamak is typically about 10 times stronger than the poloidal field. The superposition of the two fields (poloidal and toroidal) provides the helical shape (a gentle helix for the tokamak and a tightly twisted one for a pinch).

The successive tokamak devices tried out by the Soviets were successful in remedying various 'childhood diseases' of tokamaks, but in spite of these successes it cannot be said that an obviously healthy 'adolescent' was quickly emerging. The whole business was in the doldrums for quite a while and no significant progress was made with either pinch or toroidal systems (tokamaks as well as stellarators). As reported for the stellarator in the previous chapter, one of the main obstacles was that confinement seemed to be linearly proportional to the applied magnetic field strength, which was less than theoretically expected from classical considerations, which predicted a quadratic dependence. A linear dependence would require much too strong magnetic fields for fusion to become practical.

The T-3 tokamak was the first machine that managed to produce confinement that was better than linearly proportional to the applied magnetic field strength by at least a factor of ten, showing that not all magnetic confinement geometries necessarily had to suffer from this disease. These positive Russian results were first reported in 1965 at a conference in Culham (UK), but were not fully appreciated by a somewhat incredulous Western audience. After more than a decade of work, the mood in the West was very pessimistic. The 1965 Culham conference was a low point in this respect. None of the designs seemed to work. Spitzer, who reviewed the experimental work submitted to the conference, but was soon to leave the field altogether, was very pessimistic about the possibility of taming the instabilities. He stressed the, in his view, general feature that all devices in operation at that time were suffering from anomalous particle loss and that instabilities seemed to be an intrinsic property of a heated plasma in a toroidal chamber.

The Russians went home and quietly continued improving their results, so that at the next conference in Novosibirsk in August 1968, Lev Artsimovich (1909–1973), the leader of the Soviet program, could report dramatically better values for temperature, density and confinement than measured on any other device. The energy confinement times were now better by more than a factor of 50. The Soviets were not the only ones who reported improved values. Since 1965 data, some from stellarators (not the Princeton ones), but mostly from other devices, had piled up to finally discredit this linear empirical rule, so the results on the energy confinement time reported by Artsimovich were not such a surprise, but the reported temperature and density values were.

There was still doubt about these latter values, but it was laid to rest when a British team went to Moscow (at the height of the Cold War when people were dying on the streets of Prague during the suppression of the Prague Spring) to carry out an independent measurement of the temperature and density values reported by Artsimovich. They were correct. The temperature was indeed 10 million degrees. The tokamak, at that time still a device unique to the Soviet Union, had triumphed, and swept away the negative, pessimistic mood prevailing in the West.

All this had a profound effect, and indeed rightly so as it was undoubtedly the most important event in the history of fusion research so far, an event that with justice can be called a "breakthrough." The reaction, though, of people all over the world changing their research programmes overnight and scrambling to jump on the tokamak bandwagon was a little hasty, to put it mildly. Before discussing this in some detail in the next chapter, let us take a closer look at the tokamak.

Tokamak Fundamentals

a. Basic Design

Now that the tokamak has become the basic nuclear fusion device, the prime candidate for a possible future reactor, and its development will take up most of the rest of this book, it seems a good idea to summarise its basic design features, and list and discuss the fundamental parameters of both the tokamak and the plasma it is supposed to confine. This is by necessity a little technical, but that cannot be helped. Some technical details must be understood in order to know what fusion scientists are doing.

A definition of the tokamak might read: "an axially symmetric [magnetic] field configuration with closed magnetic surfaces, in which a toroidal field is produced by currents in external coils and a poloidal field by a current in the plasma. The weaker poloidal field determines the plasma confinement, and the toroidal field provides the stability." The axial symmetry referred to here is the symmetry around the z-axis in Fig. 6.1. If rotated around this axis, the configuration remains unchanged.

The basic design of the tokamak, including a transformer for inducing a current through the plasma, is depicted in Fig. 7.2. A transformer transfers electrical energy from one electrical circuit to another. A varying current in the transformer winding (primary circuit) in the figure induces a varying current in the secondary winding (Faraday's law of induction). The

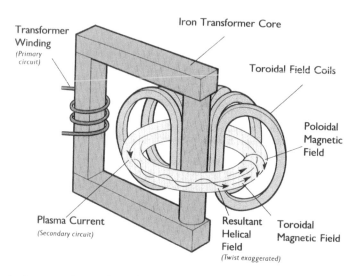

Fig. 7.2 Basic setup for a tokamak

crucial point here is that the plasma functions as the secondary winding of the transformer and that the toroidal current through the plasma (the plasma current) is generated in this way. Forget for a moment the toroidal field coils in the figure and picture just the transformer with its primary windings and the torus. When a current is sent through the primary windings, this induces a current in the torus, which acts here as the secondary winding. Apart from heating the plasma, this current will in turn generate a poloidal magnetic field around the plasma current. The combination of this latter field with the toroidal field generated by the toroidal-field coils will give the resultant helical (twisted) field.

How these magnetic fields come about is once more elucidated in the next figure (Fig. 7.3).

The drawing under (a) shows that a flowing current (through a wire for instance) generates a magnetic field around the current, while the figure under (b) shows how the circular coils produce a toroidal magnetic field that forces charged particles to travel in spirals around the torus. The figure under (c) is the analogue of figure (a) for a current flowing in a circle, like the plasma current in a tokamak, and figure (d) combines (b) and (c) to create the twisted magnetic field that we have in a tokamak. So, a tokamak is nothing else than a doughnut-shaped object (a toroid or torus and symmetric around the axis passing through the middle, i.e., axially symmetric) with a nested set of magnetic surfaces inside it that are produced by coils and currents.

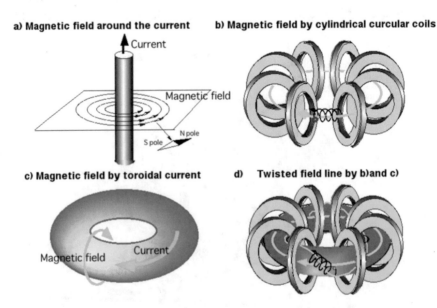

a) Magnetic field around the current

Current

Magnetic field

N pole

S pole

b) Magnetic field by cylindrical curcular coils

c) Magnetic field by toroidal current

Magnetic field   Current

d)   Twisted field line by b)and c)

Fig. 7.3 How to create a tokamak configuration

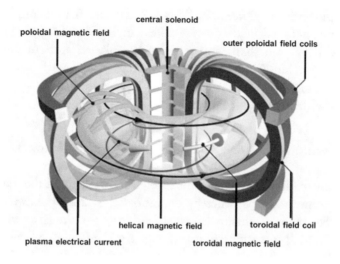

poloidal magnetic field

central solenoid

outer poloidal field coils

helical magnetic field

toroidal field coil

plasma electrical current

toroidal magnetic field

Fig. 7.4 Schematic figure of a tokamak with the various fields and currents

A more encompassing schematic picture of a basic tokamak setup has been drawn in Fig. 7.4. In this figure the transformer has been left out and some other essential elements of the tokamak are included; the central solenoid acts as the primary winding of the transformer and induces the current in the plasma. Solenoids are lengths of coiled wire that generate magnetic fields when electric current is passed through them and are responsible for driving the plasma current in all tokamak devices. The central solenoid is a long straight line of circular coils. It is the heart of the tokamak. By continuously increasing the current through the central solenoid, electromagnetic induction drives a current through the plasma. This current produces the poloidal field and heats the plasma.

The current through the central solenoid cannot of course be increased indefinitely. This is the essential reason for the inherent pulsed nature of the tokamak design. No matter how big the central solenoid coils are, a limit will eventually be reached, induction will stop, the particle drifts will jump into action and destroy the plasma confinement. A current can only be driven for a short time in this way. The conclusion is that to achieve steady-state operation in a tokamak the plasma current must be driven without use of a central solenoid to drive a current inductively. Non-inductive drive, as it is called, is needed for this.

The magnetic field coils that generate the twisted magnetic field in the torus-shaped plasma chamber must be arranged in such a way that the magnetic field lines close within the chamber and form closed magnetic surfaces to prevent particles from colliding with the walls of the vessel or from escaping in any other way. This is a delicate matter and some ring-shaped outer poloidal field coils are added in Fig. 7.4 to generate an extra vertical field that can shape and position the plasma and contribute to its stability by "pinching" it away from the walls.

b. Fundamental Tokamak Parameters

Let us now introduce some parameters that naturally follow from such a setup.

In the schematic view of the torus in Fig. 6.1 we defined the major radius R and the minor radius a, whereby the ratio R/a is called the **aspect ratio** of the torus. For the early tokamaks, aspect ratios were quite large with a being fairly small and R something like 1 m, e.g., for the famous Soviet T-3 it was 8.3 ($R = 1$ m, $a = 0.12$ m). The hole of the doughnut must be large enough for the magnetic windings (the toroidal-field coils) and the central solenoid to fit in, which limits the aspect ratio for conventional tokamaks to about 2.5. In the 1980s it was discovered that tokamaks with low aspect ratios were inherently more stable. The closer a is to R the 'fatter' the torus will start to look, like a sphere with a tube bored right through it or a cored apple.

An obviously important quantity is also the strength of the **plasma current** I_p (measured in mega amperes, millions of amperes) which produces the **poloidal magnetic field** (Fig. 7.3c), denoted by B_{pol}. It is in a plane at right angles to the toroidal field and shifts that field by a certain angle each time the **toroidal magnetic field** B_T goes around the torus the long way (i.e., toroidally). The greater the plasma current, the larger this angle.

The toroidal field B_T is typically of the order of 5 T (tesla, the unit of the magnetic field strength, equal to 10,000 gauss in formerly used units). It is typically ten times stronger than the poloidal field. For ITER the toroidal field coils are designed to produce a maximum magnetic field of 11.8 T.

The next fundamental plasma parameter is a quantity called *beta*, which is the ratio of the thermal plasma pressure (which tries to expand the plasma) to the magnetic pressure (which keeps the plasma together). It is a measure of the efficiency with which the magnetic field confines the plasma, and thus also a measure of economic merit. High *beta* is desirable but is difficult to achieve experimentally because of various plasma instabilities. For a confined plasma, *beta* is always smaller than 1 and is normally expressed as a percentage. Ideally, a magnetic confinement fusion device would want to have *beta* as close as possible to 1, as this would imply the minimum amount of magnetic force needed for confinement.

For conventional tokamaks, the record for *beta* is rather low, at just over 12%, and it is expected that practical designs would need to operate with *beta* values as high as 20%. For any given fusion reactor design, there is a limit to the *beta* it can sustain. A practical tokamak-based fusion reactor must be able to sustain a *beta* above some critical value, which is calculated to be around 5%. One of the advantages of low aspect ratio tokamaks is that they can operate at *beta* values that are significantly higher.

c. Evolution of the D-Shape

Twisted field lines are needed in toroidal devices, as we have seen, to cancel the drift in opposite directions of electrons and ions. This drift is due to the non-uniformity of the magnetic field in the torus, the fact that it is necessarily weaker on the outside of the torus compared to the inside, near the hole in the doughnut. An obvious idea to increase the plasma volume without changing the drifts is simply to extend the torus vertically into an ellipsoid, without actually changing its radius. The important point of such a D-shaped cross-section is that *beta* will be higher. In the past, various shapes have been considered, e.g., a bean-shaped cross-section and a kidney-shaped cross-section, which looked like two merged tokamaks. A D-shaped plasma is currently the most common form, and only D-shaped tokamaks are considered in modern projects. As an example, we show in Fig. 7.5 the D-shape of the plasma vessel of the German ASDEX-Upgrade tokamak.

The D-shape is not all roses though; the curvature at the corners of the D is very sharp, but fortunately occurs in only a small part of the total surface, and this part of the D can actually be usefully employed to remove waste material from the plasma, viz., the helium particles (the 'ash') of the fusion reactions. Escaping plasma is channelled into the corners of the D where special devices, called **divertors**, are placed that can handle the heavy heat load and allow for the removal of waste material from the plasma while the reactor is operating. Before discussing these devices, we first have to introduce some extra parameters to describe such a D-shaped plasma.

When the cross-section is no longer circular, we have to redefine what we mean by the minor radius and introduce other parameters, like the elongation and triangularity, to fully describe the geometry of the plasma. To know the exact form of a circle we only need to know its radius, but for an ellipse we need two parameters, and for a D-shape at least three. This is illustrated in Fig. 7.6. In the figure R is the major radius and a the minor radius, so not much has changed as far as these two parameters are concerned.

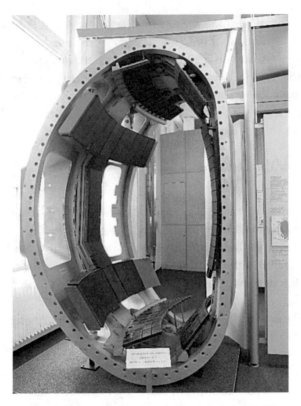

Fig. 7.5 The D-shape of the plasma vessel of the German ASDEX-Upgrade tokamak

The **elongation**, the extent to which the plasma is lengthened in the vertical direction, is indicated by the Greek letter κ and defined as the ratio of b and a: $\kappa = b/a$, where b, as indicated in Fig. 7.6, is half the height of the D. For a circular cross-section $\kappa = 1$, so the elongation measures how much the cross-section deviates from a circular form, how ellipse-like the cross-section is. The figure also shows closed field lines or surfaces (the red lines that go round in the D) and open field lines or surfaces (lines that do not close in the D). The "last closed flux surface" (LCFS) is called the separatrix (as it separates closed surfaces from open surfaces). A point in the figure at which the poloidal field has zero magnitude is called an 'X-point'. The separatrix is the magnetic flux surface that intersects with such an X-point. The figure drawn here has two separatrices. All flux surfaces external to this surface are unconfined and all flux surfaces internal to this point are closed.

Finally, we have the **triangularity**, indicated by the Greek letter δ, which measures the D-ness of the plasma shape. It is defined as the horizontal distance between the plasma major radius R and the X-point divided by

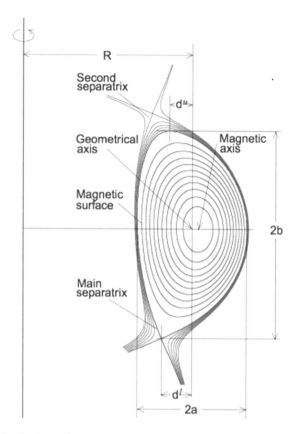

Fig. 7.6 Sketch of tokamak geometry, including the separatrix

the minor radius a ($\delta = 0$ corresponds to a fully elliptic cross-section and $\delta \sim 0.5$ to the shape of the JET tokamak, for instance). In general (when there is no up-down symmetry), there can be an upper-triangularity and a lower-triangularity as in Fig. 7.6. In this figure the distances related to the triangularity are indicated as d^u and d^l.

d. Divertors

A divertor is a device that acts as an exhaust, a sort of large ashtray that collects the 'ash' from the fusion reactions and other non-hydrogen particles. Its concept was first introduced by Spitzer for the stellarator. He suggested that the magnetic field lines at the edge of the plasma be deliberately *diverted* or led away into a separate chamber where the particles carried with them can interact with the wall and do no harm.

This is a vital part of modern tokamaks, one of its main functions being to remove impurities, e.g., from plasma-surface interactions (loosening particles from the vessel wall), and to prevent them from entering the confined plasma. It further serves to extract heat produced by the fusion reactions and to protect the surrounding walls from thermal and neutron loads. It removes power from the α-particles produced in the fusion reaction by transferring heat to a fluid, and pumps out helium ash to avoid dilution of fusion fuel. In short, it is a jack-of-all-trades as far as cleaning and heat extraction is concerned.

Early tokamak designs only very rarely included a divertor, as it was considered to be required only for operational reactors. When long-pulse reactors started to appear in the 1970s, a serious practical problem emerged. Since plasma confinement is never perfect, plasma continues to leak out of the main confinement area, strikes the walls of the reactor vessel and causes all sorts of problems. A major concern in this respect was sputtering, i.e., the ejection of microscopic particles from the wall surface, which caused ions of the wall metal to flow into the fuel and cool it. The heat load of the plasma on the wall is considerable, and although there are apparently materials that can handle this load, they are generally made of expensive heavy metals whose particles you don't want to contaminate the plasma. In the 1980s it therefore became common for devices to include a feature known as a **limiter**, which is a small ring, like a sort of washer, of a light metal that projects a short distance into the outer edge of the main plasma confinement area. It reduces the diameter of the confinement area and restricts the cross-section of the plasma column. The plasma column has to squeeze through this slightly narrower part of the tube, which keeps the plasma away from the wall. It also scrapes off the outer layer of plasma which contains most impurities. Any plasma particles leaking out will hit and erode this limiter, instead of the wall of the vessel. These erosion atoms will then mix with the fuel, and although the lighter materials of the limiter cause fewer problems than atoms from the vessel wall material, material is still being deposited into the fuel, and the more so when temperatures become higher. The limiter simply changed the source from which the contaminating material was coming and provided only a partial solution.

When D-shaped plasmas came into use, it was quickly noticed that the flux of particles escaping from the plasma could also be shaped. This resulted in the idea of using the magnetic fields to create an internal divertor that flings the heavier elements out of the fuel, typically towards the bottom of the reactor. There, a pool of liquid lithium metal is used as a sort of limiter. Magnets pull at the lower edge of the plasma to create a small region where the outer edge of the plasma, the "Scrape-Off Layer" (SOL), hits the lithium metal pool. The particles hit this lithium, are rapidly cooled, and remain in the lithium. This internal pool is much easier to cool, due to its location, and although some lithium atoms are released into the plasma, lithium's low atomic mass makes it a much smaller problem than even the lightest metals used previously.

A tokamak featuring a divertor is known as a divertor tokamak or divertor configuration tokamak. In this configuration, the particles escape through a magnetic "gap" (determined by the separatrix), which allows the energy-absorbing part of the divertor to be placed outside the plasma.

We are now all set and in the next chapters will explore how the tokamak has brought fusion further.

8

The Tokamak Stampede and Some Further Developments

With the positive Soviet tokamak results the age of despondency was suddenly over. None of the really fundamental problems, such as plasma instabilities, had been solved, but everything seemed to come together in the early 1970s. External factors, too, like Nixon's US energy independence policy, the end of the Vietnam war, and the troubles with the free supply of oil from an embargo by the Arab countries in 1973 worked very much in favour of fusion enthusiasm. This first oil crisis was followed later in the decade by a second crisis due to the decreased output in the wake of the Iranian revolution, pushing the price of oil up to about US$40. In addition, the growing concern for the environment and the clamour for clean energy, or at least energy with minimal environmental impact, meant that fusion's claim of abundant, clean energy was increasingly being heard. Even though fusion had not much to show for itself, newspapers were urging the government to support nuclear fusion research and other clean energy sources, all in the completely mistaken notion that a nuclear fusion reactor was just around the corner. Very little was actually known about the environmental, biological and safety issues of fusion reactors, for the simple reason that nothing was working yet. No fusion power had been produced. It was not even clear that a well-behaved plasma could be created and confined long enough for any power to be produced. It is still not clear today!

Because of the increased enthusiasm among scientists after the breakthrough with the tokamak, matched by increasingly generous government funding, the 1970s became the glory years of fusion. However, they were not

L. J. Reinders, *Sun in a Bottle?... Pie in the Sky!*, https://doi.org/10.1007/978-3-030-74734-3_8

glorious because of any great strides towards energy from nuclear fusion. The T-3 tokamak results reported at Novosibirsk unleashed a veritable tokamak stampede. Some forty tokamaks were built in this decade in the US, Europe (France, Germany, Italy, UK), Japan and the Soviet Union. The website http://www.tokamak.info/ gives an (incomplete) list of some 185 conventional (i.e., excluding spherical) tokamaks, that have been built till the present day all over the world. These devices have all contributed something to the great fusion database. In the following we will only mention those that have had some real importance or remarkable results.

A further change that happened at about the same time in the American fusion effort was that strategic decision-making was taken away from the individual laboratories. Funding had always come from the AEC, but the various laboratories had decided themselves which problems to study and which machines to build. They had a tendency to go ahead with their own ideas even if the AEC did not approve, confident that things could be put right later. This changed when Robert L. Hirsch (b. 1935) became director of the AEC fusion program in 1971. He pursued a much more aggressive policy and reduced the independence of the lab directors, moving the centre of control and policy-making to Washington. He committed the fusion community to a program of milestones and ever larger machines, steering them away from the fundamental research program of the 1960s towards the practical goal of creating *commercial* energy from fusion.

This momentous and premature change of direction from mostly pure research to constructing a working fusion reactor had a considerable impact on the research topics studied in the various research groups. The autonomous development of a scientific field, in this case plasma physics, is disturbed when a practical goal like a fusion reactor is formulated. It steers the research in a certain direction and restricts the problems that must be solved and the order in which this should be done. It determines when a certain development must be broken off, namely when it is considered to distract too much from the practical goal; and it also determines what must be considered as a solution and when the solution is considered adequate. Pure plasma physics had to align itself with the practicalities of constructing a fusion reactor based on the tokamak design, and this at the time when plasma physics was still struggling to make any sensible predictions about the behaviour of plasmas in general, let alone in such a complicated geometry as a toroidal magnetic system.

Hirsch took the distribution of money in hand, put pressure on labs to do his bidding, and sparked off competition between the various labs for the funds needed to build ever larger machines. He was a good salesman and in 1971 told Congress that with vigorous funding a demonstration fusion reactor could be ready by 1995. A totally unfounded prediction as the recent Russian tokamak results in no way justified such a pronouncement and not a single tokamak was yet running in the US or anywhere else outside the Soviet Union. In a later report the claim was even upped: "*an orderly aggressive program might provide commercial fusion power about the year 2000, so that fusion could then have a significant impact on electrical power production by the year 2020.*"

Hirsch very much exemplifies what is wrong with the approach to fusion. At one point he is quoted as saying: "*I had inherited a collection of very bright people, very good physicists, who in many respects were not aggressive but who were very conservative, and wanted to do a lot more basic physics than I thought appropriate. They wanted to solve all the problems before they would take the next step. I had a bunch of people who were generally timid and only weakly committed to making practical fusion power.*" Hirsch apparently did not understand that commitment is not sufficient for solving a problem. He was of the opinion that increasing the size of the tokamak would move the walls further away from the plasma, which would then take longer to leak away and that all problems would thus be solved. Without further ado he wanted to build large tokamaks and straightaway burn deuterium and tritium, with all the complexities of handling radioactive tritium and without proper preparations. Not the way to do science, one would think. But all this bravado got him the money he wanted, and during Hirsch's stewardship, from 1972 to 1976, funding for fusion in the US leaped from $30 million to $200 million.

In the wake of the tokamak stampede a variety of research directions were abandoned in the US in the course of a few months, and years of research experience were thrown out of the window in favour of the blinkered pursuit of the tokamak concept, on the flimsy basis of a single Russian result that had not yet been independently confirmed on any other device. The change into an energy generating project, perhaps understandable in the general euphoria, came just a few years later.

In December 1969, within two months after confirmation of the Russian results by the British, a drastic switch to tokamaks was made at the Princeton Plasma Physics Laboratory (PPPL). Operation of the Model-C stellarator was ceased and the device converted into a tokamak, renamed Symmetric Tokamak. Within a three-year period, from 1970 to 1973, the Princeton team entirely shifted its focus away from the stellarator and built a series of ever larger tokamaks, culminating in the $300 million Tokamak Fusion Test Reactor (TFTR) to be discussed in the next chapter.

It was a surprising decision and almost certainly premature, as Model C was still a promising machine and numerous experiments had yet to be completed. The stellarator had been studied for 18 years and a lot of expertise and knowledge had been gained, which would now for the most part go to waste.

The fusion program emerging now in the US was completely unbalanced, with the various laboratories competing for the funds, albeit soon to be increased. Duplication was inevitable, and the great emphasis on tokamak research created sky-high expectations for a major breakthrough. After all, if everyone in the fusion community suddenly caught tokamak fever, than surely that must pay off. Or else, what were they doing?

There was also a grave political aspect to this business, causing great worries in Washington. The Soviets were planning a whole series of tokamaks, culminating in the very large T-10 tokamak, which they expected might be capable of reaching full reactor plasma conditions. That would once again (after the Sputnik and the first man in space) snatch the glory away from the United States for being the first to demonstrate the scientific feasibility of controlled nuclear fusion. They needn't have worried. Even today that time has still not come and the Soviet Union has in the meantime already been buried.

For science there was a great and a rather unexpected advantage in all this. Now that national pride was at stake, the members of the Congressional oversight committee were suddenly 'unusually receptive' and even asked what the level of support for the fusion programme should be. No wonder that the proposals for tokamaks were coming in fast, even before the Soviet results had been confirmed, e.g., from the private company General Atomic, Oak Ridge National Laboratory (ORNL), MIT, and University of Texas at Austin.

At Princeton the first US tokamak results, comparable to the T-3 as regards confinement times and temperatures, but not remarkable in any other respect, were already presented in July 1970, while Oak Ridge and MIT took until late 1971 and early 1972, respectively, to produce their first results.

Especially at Princeton, funds were set aside for various follow-up devices of which the Princeton Large Torus (PLT) is especially noteworthy. It was an ambitious experiment that was designed to produce higher plasma currents than the Russian T-3 and was largely a copy of the Russian T-10, but with novel external heating systems. It was the first tokamak which achieved a plasma current of 1 megaamperes (MA) and managed to push the ion temperature above 5 keV (60 million degrees). The latter achievement, accomplished in 1978, is important as it was the first time that the ion temperature exceeded the critical threshold for ignition, i.e., the temperature at which the nuclear fusion reaction can become self-sustaining. With the overstatement so common in fusion research, the American fusion community considered this the most significant achievement in the then nearly 30 year history of fusion research, but for some reason they forgot to call it a breakthrough. A worrying concern was that confinement times seemed to get worse when the neutral-beam heating was increased, and this cast a small dark cloud over the future of the big TFTR that was already being planned at that time.

In 1981, more than ten years had passed since the tokamak stampede had begun and, apart from a few 'world' records in temperature, and 'firsts' in the deployment of various techniques, not much had been achieved, while according to the PPPL website, PLT experiments had been expected "…*to give a clear indication whether the tokamak concept plus auxiliary heating can form a basis for a future fusion reactor.*" The website fails to answer whether the expectation was fulfilled. Well, it was not, but in spite of this, PLT was still considered an extremely successful experiment and led to a machine, TFTR, allegedly capable of reaching breakeven, which ushered in the age of the big tokamaks.

It makes little sense to discuss here in any detail the various machines built by other laboratories in the US, and I just mention that ORNL, MIT, and General Atomic all contributed to the tokamak frenzy by proposing and building various devices that did useful work in various areas of tokamak research, but eventually lost out to PPPL which became and still is the main US nuclear fusion laboratory.

In Chap. 5 we mentioned the phenomenon of disruption, an event in which all the energy is dumped suddenly in a few milliseconds into the structure of the machine. It is a serious problem and a major hazard for any tokamak. Unless it can be solved the structure of the tokamak, and especially the divertors, have to be beefed up to be able to absorb all the energy, adding considerably to the construction costs. Normally the plasma energy escapes slowly into the divertors, which are designed to handle such heat loads. The MIT group working with one of their tokamaks, called Alcator C-Mod, was able to capture what happens to a plasma in a disruption. This is illustrated in Fig. 8.1. In such a typical elongated D-shaped tokamak, specially shaped coils must prevent the plasma from drifting up or down. When an instability causes a disruption, the plasma moves vertically, shrinking as it loses its energy. In the figure it moves downward towards the divertor, but it could also move upwards. The time scale shows that the whole event, which meant the end of the plasma, took just 6 ms. So, if you want to avoid a disruption this is the time scale in which to act, just 6 ms to spot the onset of the disruption and take action to prevent any serious consequences.

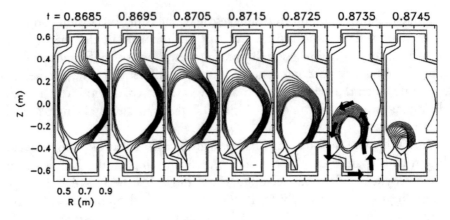

Fig. 8.1 Vertical motion of the plasma in a disruption

General Atomic was one of the first to develop D-shaped tokamaks, a shape that is now widely used in what are called 'advanced tokamaks'. They are called advanced as they are characterised by operation at high plasma *beta*, with strong plasma shaping and active control of various plasma instabilities. Their latest model, DIII-D, is still operating and currently the largest tokamak facility in use in the US. It holds the record for *beta* (12.5%) for conventional tokamaks.

In Europe various countries took equally bold steps. The British at Culham who were in the middle of the construction of their own stellarator, now decided to leave out the helical windings and to start operating it (in 1972–1973) as a tokamak, later followed by a purpose-built tokamak.

For the French, too, 1968 was a watershed year. They now suddenly knew what to do and without further ado embarked on the construction of what was for that time a large tokamak, the TFR (Tokamak de Fontenay-aux-Roses), similar in size to the Soviet T-3, but with a higher toroidal magnetic field. It was followed by the Tore Supra, one of the first tokamaks with superconducting toroidal magnets. Superconductors have zero resistance, so once a current has been started in them, it will keep running (almost) forever. It must in the end provide the solution to the still elusive steady-state operation of the tokamak. The drawback is that these superconductors have to be cooled to below 4.2 degrees Kelvin (about –269 °C). This cooling consumes enormous amounts of energy and is a serious impediment to making the generation of fusion energy even possible, let alone commercially attractive. An important development is that present-day superconducting technology allows magnets to be made out of materials that become superconducting at higher temperatures, reducing energy consumption for cooling.

When Tore Supra had fulfilled its mission in 2011, it was upgraded to a device called WEST (which stands for Tungsten (chemical symbol 'W' for Wolfram) Environment in Steady-state Tokamak. The acronym WEST was specifically chosen to set it against the Chinese tokamak EAST. Both WEST and EAST are part of SIFFER, the Sino-French Fusion Energy Center (see below). WEST involved conversion to a D-shaped cross-section and the installation of a tungsten wall for the plasma vessel and a divertor. It has been in operation since 2016.

In Germany the IPP showed mild symptoms of tokamak fever, starting at a smaller scale before committing itself to larger machines. Stellarator research was not abandoned and since that time IPP is the world's only institute pursuing research on both tokamaks and stellarators.

It was on one of the German tokamaks, the ASDEX (Axially Symmetric Divertor Experiment) operated at IPP in Garching, that a momentous discovery was made in the early 1980s, essentially the first really important one in the age of the tokamak, and we must discuss this in some detail.

H-Mode and ELMs

When discussing the experiments with PLT at Princeton, it was mentioned that confinement times got worse when the plasma temperature was increased by external heating. The problem was completely unexpected and scientists were at a loss about its cause. According to new calculations gigantic amounts of power would be needed to heat the plasma. Under such circumstances it seemed impossible to ever achieve a burning plasma. The solution, as unexpected as the problem itself, came in 1982. It was found completely accidentally on the German ASDEX tokamak that, when increasing the heating power slightly from 1.6 to 1.9 MW, there was an abrupt transition in the plasma characteristics; the plasma snapped into a new mode as it were. Confinement improved by about a factor of 2, which seems little considering that confinement times have increased millionfold since the beginning of fusion research, but it may be the last push needed to get to the regime of a proper reactor. The discovery of this high-confinement regime in the ASDEX tokamak, for short H-regime or H-mode, versus L-mode or low-confinement mode which denotes the normal confinement in tokamaks, is one of the most important discoveries in fusion research. Essentially only two achievements in fusion are worth the word "breakthrough": the Russian result with the T-3 tokamak in 1968 and this H-mode discovery in 1982.

H-mode performance is achieved via the spontaneous formation of a transport barrier in the outer few percent of the confined plasma, as the result of suppression of the turbulence in the few centimetres of plasma just inside the last closed flux surface.

The pioneering achievement of H-mode had to do with the use of a divertor, or actually with the presence of a separatrix (see Chap. 7). The separatrix signals the transition from the confined plasma with closed field lines to the Scrape-Off Layer (SOL) with open field lines. The boundary layer produced by the divertor induces a transport barrier at the plasma edge, creating good thermal insulation. Divertor operation and the H-regime have now become the standard in modern tokamaks, right up to ITER.

In the following years, H-mode was observed in several other tokamaks. In 1993 H-mode was also achieved in the German Wendelstein 7-AS stellarator, thus demonstrating that it was a "generic feature" of all toroidal configurations and can indeed be reproduced in any tokamak or stellarator. For the H-mode to arise it seems that only two requirements have to be met: (1) that the input power be high enough (the actual level varies from device to device); and (2) that the plasma be led out by a divertor into an external chamber rather than allowed to strike the wall.

The physics behind the H-mode is still not completely understood. Such lack of understanding is a recurring feature in fusion research and has never greatly bothered fusion scientists. The value of H-mode is unquestioned. All present-day tokamaks are designed to run in H-mode. It was vital for JET in achieving its record results in the 1990s (see Chap. 9) and without it ITER would need to be twice as large and, consequently, twice as expensive.

But, as usual, nothing comes for free, and while the plasma characteristics in the H-mode are better than in L-mode, a new type of instability emerged. These are the edge-localised modes (ELMs), already mentioned in Chap. 5. They occur most violently in a thin layer of the plasma in the edge region of the torus and pose a major challenge for nuclear fusion with tokamaks.

Not surprisingly, ASDEX was also the first tokamak on which these very troublesome ELMs were seen. An ELM is a periodic distortion of the plasma boundary, which rotates with a velocity of several kilometres per second and exists for about half a millisecond, ejecting bundled plasma particles and energies outwards to the vessel wall. It appears in the standard H-mode operation regime and leads to a crash ('ELM crash'), causing plasma to eject particles in bursts onto the chamber wall. Up to one-tenth of the total energy content can thus be ejected. In bigger future devices the particle bursts caused by the ELMs may lead to the destruction of plasma facing components (PFCs) and can cause overloading of, in particular, the divertor. Their mitigation is therefore of paramount importance.

The Soviet Program

The Soviets did of course not have to change course and continued their program at a great pace with the T-4, T-5, T-6, etc., ending with the T-15, whose construction was started just when the Soviet Union was in its death throes in 1988. In their eagerness to be the first to prove scientific feasibility the Soviets designed ever larger tokamaks, culminating in the design of the giant T-20, a genuine test reactor. In the end they sensibly settled for a more modest, but superconducting physics experiment, the T-15. The T-20 has never been mentioned again. It was all based on the idea that larger is better: when you scale up the reactor size, e.g., increase the plasma radius by a factor of two, confinement times would improve fourfold. Physicists call this scaling laws, where the use of the word 'law' is somewhat pretentious for something that is essentially just an empirical rule. This is also the idea behind building the giant ITER facility. The trouble is that these scaling laws, being just experimentally observed patterns or regularities, do not always work, or the scaling just stops at a certain point. Lacking a thorough theoretical understanding, it is, however, the only thing available.

In general, Soviet research gives the impression of being conducted in a careful and well-thought-out manner within the framework of a long-term program, more so than in other countries, in particular the US, and in the way you expect science to be conducted, although some rigidity is also apparent. No competition for funds between the various institutions, although there must have been some rivalry; no jumping from one extreme to the other, cutting funding one year and increasing it again the next. And although they took notice of what was going on in other countries, it did not divert them from their course. In the view of Lev Artsimovich the true danger of international contacts for the domestic Soviet program was that physicists would pay too much attention to foreign programs and mistakenly go down the wrong path, chosen by others, rather than finding their own way. That is why he stubbornly stuck to tokamaks up to his premature death in 1973.

The Soviets were also putting a lot of money and manpower into nuclear fusion. They led the world with twice the US effort, as well as in the number and variety of major experimental devices. In plasma theory the Soviets were also preeminent, about four times the US effort. In spite of this, judging from the list of notable discoveries and progress with tokamaks from http://www.tokamak.info/, they lost their edge on the rest of the world fairly soon after the stampede began. The last notable discovery or 'first' with a Soviet tokamak dates from the early 1970s with the T-7, which was the first tokamak to use superconducting magnets.

It has been suggested that the main obstacle to further Soviet success was the big lag in computing power, diagnostics and modelling. Another reason may have been that Artsimovich, their greatest specialist in the field, died prematurely at the age of 64 in 1973. Evgeny Velikhov (b. 1935), who upon Artsimovich's death took over the leadership of the Soviet fusion program, was a much less dynamic leader. Moreover he was a theoretical plasma physicist and it is in general not a good idea to place a theoretical physicist at the helm of a largely experimental and engineering project. Unless the direction of the project is clear, which is by no means the case for nuclear fusion, it is bound to veer off in one way or another or get into a rut.

Another possibility for the lack of any major results from Soviet/Russian tokamaks is of course that the tokamak is after all not the ideal design for a fusion reactor, but it seems that this possibility is not really in the minds of any fusion enthusiasts.

Notable in the Soviet series is the T-10, which started operating in 1975. It is one of the few larger tokamaks of that era still in use. The T-10 was supposed to be shut down around 1985, but when it became clear in 1983–1984 that the construction of the new big tokamak T-15 would be delayed for four to five years, it was decided to preserve the T-10, as otherwise there would be no sizable tokamak left to do experiments with. It has been around ever since. The physics for the project had already been formulated in 1968. The Princeton Large Torus was largely a copy of the T-10, but had superior peripherals, like heating systems and computers for handling data. At the time T-10 was the biggest tokamak in the world, but it was poorly conceived from the start and remained a machine of limited capabilities.

Research on D-shaped plasmas also originated in the Soviet Union where Artsimovich and Shafranov noted in 1972 that, in a tokamak with a vertically elongated plasma cross-section (a D-shaped plasma), stability conditions were better than with a circular cross-section, and that consequently it should be possible to obtain higher plasma temperatures. The elongated cross-section also seemed to promise higher *beta* values. It was put into practice in 1973 on the relatively small T-9 tokamak, which had an extremely elongated cross-section.

In the course of the 1970s and 1980s the times for funding fusion became increasingly unfavourable, because of financial pressures connected with the construction of nuclear fission reactors. Moreover, as in the US, the optimistic, but never fulfilled promises of the fusion scientists made government agencies sceptical. In spite of this, in June 1983 the Politburo still increased funding in line with the Soviet Union's long-term energy plan that envisaged building industrial tokamaks early in the twenty-first century. That vision was at least a century off.

In line with this, tokamak research at the Kurchatov Institute developed further with the introduction of the T-15, designed for the production and analysis of a plasma with thermonuclear parameters, a goal that was not realised either. The T-15 was also intended to secure the Soviet Union's global leadership in fusion research once again, an intention that likewise failed miserably. Contrary to other large tokamaks designed and built from the early 1980s onwards (like JET), the T-15 still had a circular cross-section. It is surprising that of all the tokamaks built at the Kurchatov Institute only the T-8 and T-9 (still built under Artsimovich's supervision in the early 1970s) had a non-circular cross-section. The T-15 was optimistically called the first industrial prototype fusion reactor and used superconducting magnets to control the plasma. These enormous magnets confined the plasma, but failed to sustain it for more than a few seconds. The USSR's desire for cheaper energy ensured the continuing progress of the T-15 in the twilight years of the Soviet Union under Mikhail Gorbachev. Reactors based on the T-15 were supposed to replace the country's use of gas and coal as the primary sources of energy. Needless to say, there is no chance of this happening any time soon. In the Soviet Union, too, fusion scientists were utterly devoid of any sense of realism as regards the feasibility of energy from nuclear fusion.

The T-15 achieved its first thermonuclear plasma in 1988 and the reactor remained operational until 1995, but during this entire period it was plagued by technical and budgetary problems, undoubtedly connected with the demise of the Soviet empire. So, all in all, it can be concluded that the T-15 has largely been a waste of time, energy and money.

But, things could have been worse, since the planning in 1983 (it was already talked about in the mid-1970s) had at first foreseen, in the construction of the T-20, a very large tokamak with superconducting coils that would greatly surpass the big tokamaks to be discussed in the next chapter. It would have been a giant machine of close to ITER proportions and twice as large as JET, but (surprisingly) with a circular plasma cross-section. It is bewildering to read in a 1975 conference report that: "*The parameters of the T-20 are determined by its purpose, i.e., the possibility of prolonged operation with a reacting D-T plasma and the technical possibility of attaining these parameters in the early*

1980s." Now almost fifty years later this situation has still not been achieved with any tokamak, and will not, for a considerable time to come.

From its early successes in tokamak research in the 1950s and 1960s, the brightly burning flame of Soviet fusion research has become duller and duller and was reduced to a flicker at the collapse of the Soviet Union. Russia took over the T-15 programme from the defunct Soviet Union and the T-15 is expected to remain the main tokamak in Russia to the year 2040, when the country should have been studded with 'industrial tokamaks'.

Developments in Asia

In Japan the programme of nuclear fusion projects has been very extensive. Fusion as a future source of energy apparently was and is taken very seriously in Japan, perhaps understandable for a country that has very few fossil fuels of its own (apart from coal). Research in plasma physics and nuclear fusion began in Japan around 1955 and instead of immediately starting the construction of relatively large-scale fusion experiments, the Japanese very sensibly decided to start first with basic plasma studies through the construction of small-scale devices and the training of young scientists at universities.

Late in the 1960s, after the Russian T-3 results, construction of fusion devices really took off and Japan very swiftly jumped on the tokamak band-wagon. Its fusion efforts entered a second stage through the launch of a series of toroidal plasma projects and various pinch experiments. This was further expanded in the mid-1970s by authorizing the construction of a large-tokamak (the JT-60), for carrying out a scientific feasibility experiment as a milestone towards the realization of fusion energy. The JT-60 is one of the big tokamaks to be discussed with JET and TFTR in the next chapter.

More than 40 conventional tokamaks have been built in Japan at various institutions from the 1970s onwards, more than in any other country, in addition to 10 spherical tokamaks and a large number of devices of alternative design. Of the 40 conventional tokamaks only a couple (two or three) are still in operation.

Japan apparently has great confidence in the prospects for fusion energy and is already looking beyond ITER. After the 2011 Fukushima disaster it was decided to gradually phase out power generation by nuclear fission. Plans were drawn up to shut down all its nuclear power plants and the construction of new plants was cancelled in the face of strong opposition. It is extremely doubtful that this is a good policy to follow, as sizable power generation from nuclear fusion is still but a distant and dim prospect. No wonder that these plans have been revised and the target is now to continue generating 20–22% of total power from nuclear fission up to 2030. The fact is though that Japan is taking a leading role in developing nuclear fusion as a next-generation power source and is cooperating closely with Europe in this respect. We are running ahead a little now, but within the framework of ITER, Europe and Japan have identified several major projects to be carried out jointly in Japan. These include the construction of a materials test facility (International Fusion Materials Irradiation Facility (IFMIF)), and the construction and operation of an advanced superconducting tokamak, the JT-60SA, the successor of the JT-60 and to be discussed in the next chapter. The IFMIF is urgently needed for developing suitable materials that can withstand the onslaught of the neutrons released in the nuclear fusion reactions.

South Korea, which relies on imports for more than 90% of its energy needs, has been a player in nuclear fusion since the oil crisis of the 1970s when it developed its first tokamak. In 1995, the Korean government launched KSTAR, the Korea Superconducting Tokamak Advanced Research, the flagship of South Korea's research in nuclear fusion. It is a medium-sized all superconducting tokamak, one of the first research tokamaks in the world featuring fully superconducting magnets. It will study aspects of magnetic confinement fusion that will be pertinent to the ITER fusion project, which South Korea joined in 2006. In 2016 KSTAR achieved a high-performance H-mode plasma that was stable for 70 s, a world record at the time and since broken by the Chinese EAST.

India has a small, indigenously designed and fabricated tokamak, running since 1989 and used as a testbed for the SST-1 (Steady-state Superconducting Tokamak). The SST-1 started in 1994 and achieved first plasma in June 2013. It is part of a new generation of advanced tokamaks with a D-shaped plasma and steady-state operation (i.e., pulses up to 1000 s) as its major objective. It has been designed as a medium-sized tokamak with superconducting magnets. Although the SST-1 is not yet in full operation, its successor the SST-2 is already being designed and a demonstration reactor contemplated. This seems an established practice in the fusion business, where even close to 100 years of failing research has not yet instilled a sense of patience and reflection into the community.

In April 2019 Reuters reported that "China aims to complete and start generating power from an experimental nuclear fusion reactor by around 2040", again reverting to the magic two decades that always separate us from fusion power. China is part of the ITER consortium. Everything depends on ITER's success and it will probably already need unparalleled luck to get

ITER properly working by 2040, as will be discussed in detail in Chap. 10. Of course, optimism is a good thing, but failing to learn from the past (and the present) can be fatal. Reuters was not the only one to sing the praises of China's nuclear fusion prowess in April 2019. There were stories about "Earth's Second Sun—China's Fusion Future, the Holy Grail of Unlimited Energy" and even the BBC joined in with the less exalted heading: "Will China beat the world to nuclear fusion and clean energy?" The press apparently never learns to take such stories with a large pinch of salt. Are they really unaware of the fact that scientists have been telling such stories now for close on seventy years, with the epithets used becoming ever more exalted?

It is of course not surprising that China puts great store on nuclear fusion as a possible future source of energy. Like Japan, the country is heavily dependent on foreign suppliers for oil and gas to meet its voracious energy needs (although it produces a lot of coal, which provides about 60% of Chinese energy, and natural gas of its own). Power from nuclear fission has also been meeting opposition in China, since the Fukushima disaster. As of March 2019, China has 46 nuclear power stations in operation with a combined installed electric capacity of more than 45 GWe (gigawatt electricity). Nuclear power contributed about 5% of total Chinese electricity production in 2019. Another 11 nuclear reactors with 11 GWe of electric power will be added over the next two years. Additional reactors are planned for a further 36 GWe. China was planning to have 58 GWe of capacity by 2020. However, few plants have commenced construction since 2015, and it is now unlikely that this target will be met.

China's population is huge, 1.4 billion at present and forecast to grow to 1.5–1.6 billion by 2050. While energy consumption per capita in China is at present still fairly low, it is expected to grow rapidly. However, this does not necessarily have to be the case as the example of the UK shows. Today the UK consumes less energy than it did in 1970, despite an extra 6.5 million people living there. Per capita its consumption is 2764 kg of oil equivalent, against a figure for China of 2237, so not all that different, surprisingly close actually. In 1970 this figure was still close to 4000 kg of oil equivalent per capita for the UK, staying almost constant until 1995, after which it started to decline rapidly to reach the quoted figure (of 2015). It seems that the UK is more efficient both in producing energy and using it. Households use 12% less, while industry uses a massive 60% less. This is largely offset by a 50% rise in energy use in the transport sector. Still the amount of energy used per capita in the UK is 30% less than in 1970! This example shows that China can perhaps gain more by introducing energy savings and efficiency schemes than by investing heavily in nuclear fusion monsters. This is also borne out by the fact that the efficiency of turning fossil fuel (coal) into economic output in China is very poor, just one seventh of that in Japan.

Experimental plasma physics research in China began on linear pinches by several groups in Beijing in 1958, but has developed extensively, especially since the 1970s, when various tokamak devices were constructed.

Since 2017 the Chinese have collaborated extensively with the French in the Sino-French Fusion Energy Center (SIFFER). EAST, the Experimental Advanced Superconducting Tokamak, is one of the devices brought into SIFFER by China (WEST being the French contribution, see above). EAST is the first non-circular advanced steady-state fully superconducting tokamak. Construction started in October 2000 and was completed in March 2006. So it took only 5 years to build and the claim is that costs were a mere $37 million. EAST is designed on the basis of the latest tokamak achievements of the last century. Its distinct features are a non-circular cross-section, fully superconducting magnets and fully actively water-cooled plasma-facing components. Its construction and physics research will provide direct experience for the construction of ITER. It is smaller than ITER, but similar in shape and equilibrium, yet more flexible.

9

The Big Tokamaks: TFTR, JET, JT-60

Tokamaks are getting bigger and bigger, with ever larger major radius R, and also use an increasingly stronger confining magnetic field. Why is that necessary? The simple reason is that the bigger the machine the more time the particles spend inside the plasma before leaving it, and the more time they have to fuse and produce energy. A particle in a plasma gyrates around a magnetic field line while moving along this field line, and sometimes collides with another particle. In such a collision, the guiding centre of its motion will jump two gyro-radii (see Fig. 5.2 to remind yourself of this motion), so every collision causes the particle to jump towards the wall. Such collisions are necessary, as particles that do not collide will of course never fuse. In a fusion reactor a deuterium nucleus will typically have a gyro-radius of 1 cm. It takes more than 1000 collisions before a fusion reaction occurs. In these 1000 collisions the particle may come ever closer to the wall, unless the machine is big enough to allow for so many collisions. In other words, the bigger the plasma volume the more energy will be generated, while losses are proportional to the surface area. In the following we will look at this in a little more detail.

Due to the turbulence of the plasma, it is still impossible to calculate energy and particle transport in tokamaks from first principles. Different regimes of turbulence are responsible for different aspects of plasma transport, e.g., particle diffusion and heat conduction, in different regions of the plasma, whereby these various turbulence regimes interact with and influence each other. Energy losses through electrons, for instance, exceed those predicted

L. J. Reinders, *Sun in a Bottle?... Pie in the Sky!*, https://doi.org/10.1007/978-3-030-74734-3_9

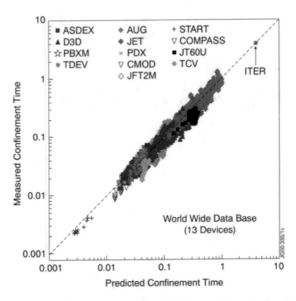

Fig. 9.1 Showing the agreement between experimental results for the confinement time from a number of different machines (ordinate) and the result of the scaling law (abscissa) extrapolated to ITER

by theory by a factor of ten. Many real loss mechanisms are not clear. In the absence of adequate theoretical understanding, empirical methods have to be relied upon. That means collecting data from many tokamaks over a range of different conditions and parameters and trying to establish from them how the confinement depends on the plasma parameters, and subsequently extrapolating the results to higher values of these parameters. From such extrapolations one can then derive how a new machine should be constructed. As mentioned before, such extrapolation relationships are generally called 'scaling laws'. They are not really laws, but no more than empirical rules of thumb, e.g., connecting confinement times with machine and plasma parameters. Figure 9.1 shows an important and much exhibited figure in this respect: the empirical scaling of energy confinement time from present day experiments to ITER. Due to the logarithmic scales chosen (both for the ordinate and the abscissa) it looks better than it actually is. A log–log plot as in Fig. 9.1 is a great way to make a relation look linear. Moreover, it only has data from 13 devices, hardly a large sample.

From this, it is concluded that the energy confinement time of tokamak plasmas scales positively with plasma size (meaning that it increases with plasma size). It is consequently generally expected that Lawson's triple

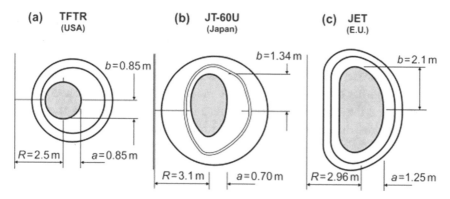

Fig. 9.2 Comparison of the three large tokamaks TFTR, JT-60U and JET; the size of the fusion chamber and the sizes of the plasma volume for each device

product will also increase with plasma size. This has been part of the motivation for building ever larger devices, up to ITER with a plasma volume of over 800 m³.

However, tokamak plasmas are also subject to operational limits. Two important limits are a density limit and a *beta* limit, *beta* being the ratio of the plasma pressure to the magnetic pressure (see Chap. 7). Such limits have to be obeyed to avoid the development of instabilities and disruptions. Current and density limits are usually determined by disruptions, while other instabilities set the *beta* limit.

In this chapter we will discuss three tokamaks that are normally called the big tokamaks: the American Tokamak Fusion Test Reactor at PPPL in Princeton, the Joint European Torus (JET) at Culham in the UK, and the Japanese JT-60 (and its upgrades JT-60U and JT-60SA). They link the small and medium size tokamaks of the 1970s and 1980s to a thermonuclear experimental reactor like ITER. Figure 9.2 compares the cross-sections of the three devices.

Tokamak Fusion Test Reactor (TFTR)

The idea for TFTR goes back to the very early 1970s. The fact that it was conceived at precisely this early, not to say premature, stage in tokamak research, was mainly due to Robert Hirsch, and perhaps also due to pressure from the decision taken by the European Union in 1971 in favour of a robust European fusion program. From the onset of his term in office starting in 1971, Hirsch was bent on building a tokamak feasibility experiment, i.e., an experiment that would produce and contain a high temperature plasma exceeding the Lawson criterion. He wanted to make a major step towards future power generation, so he insisted on a machine that would burn deuterium and tritium, the most promising fuel combination for achieving fusion. Of course, there was widespread expectation in the fusion community that one day that step would have to be made, but the general consensus was that the time for this had not yet come. Hirsch had other ideas though. In the deuterium-tritium (D-T) reaction, shown below

$$D + T \rightarrow {}^4He \ (3.5 \ MeV) + n \ (14.1 \ MeV)$$

the energy of the α-particles or helium-4 nuclei (^{4}He) at 3.5 MeV or 3500 keV is roughly 300 times the 10 keV (in temperature about 100 million degrees) of the deuterium and tritium fuel ions. Being charged, they are captured by the magnetic fields, remain in the plasma and substantially alter its properties. Hirsch felt that this had to be studied as it would be a typical situation for a functioning fusion reactor. In principle there was not much wrong with this idea; the only snag was that at that time nobody knew how to get a burning plasma that would produce these α-particles. In his view this lack of knowledge was essentially due to the attitude of the fusion scientists. He wanted them to change their attitude from a scientific plasma physics interest (to find out how it might work) to the practical goal of energy generation (it doesn't matter how it works, so long as it works), in other words he wanted to turn them from scientists into engineers. A D-T experiment would provide the opportunity to automatically shift the attention to engineering problems, as Hirsch was of the opinion that that was where the real problems lay. And finally there would be the political advantage, it would bring home the importance of fusion to the public and the politicians by demonstrating the actual production of power. The amazing thing in all this is that, although most physicists and engineers did not like his suggestion, nobody managed to stop him. The physicists objected as the physics of tokamaks, so early in the game, was still very poorly understood, while the engineers abhorred a D-T burning tokamak as it would be plagued with problems of radioactivity (tritium is radioactive and easily absorbed in the human body, implying that much of the work must be done by remote handling) and hazardous neutron fluxes (the high-energy (14.1 MeV) neutrons that are also released in the D-T reactions would make the apparatus radioactive). But the fusion community was divided, a lot of money was being offered for doing this experiment, and especially Oak Ridge, which had a long history of dealing with radioactive reactors, was positive. They thought they could build such a tokamak experiment at reasonable size and expense. By feeding them money Hirsch got them to design the tokamak he wanted and went even further by plumping for a device that would not just achieve breakeven, but even ignition.

Princeton, also a prime candidate for this venture, had great misgivings because of the use of radioactive fuel. It was not a federal laboratory, but a university, where they felt radioactivity had no place. The Princeton laboratory saw itself purely as a plasma physics laboratory, in contrast to Oak Ridge which was and always had been a reactor laboratory. The scientists at Princeton and many others in the fusion community were not convinced either that the scientific problems connected with the burning of deuterium and tritium were the important ones. At this point in the fusion game there were far more important problems to study than the burning of deuterium and tritium, such as impurities and diffusion losses.

Apart from the radioactivity aspect, Princeton was as eager as Oak Ridge to build a tokamak feasibility experiment, the successor to their PLT experiment, for which funding had just been obtained. Hirsch at first favoured Oak Ridge as he was very encouraged by recent results they had obtained on neutral-beam heating and, due to his engineering background, the technical orientation of their work was in general more to his taste than Princeton's physics approach. However, the expected doubling or tripling of ion temperatures at Oak Ridge with neutral-beam heating turned out to be elusive and Princeton scooped them also in that respect. On top of that, the Oak Ridge design became too costly at several times the hundred million dollars budgeted for the experiment.

The competition between the two laboratories went on for some time. Both started to design a D-T burning experiment. Although Princeton still had misgivings, it did not want to lose the feasibility experiment. Refusal might jeopardize its future. The outcome of the rivalry was such that the prospects dramatically reversed in the course of a single year. In September 1973 Oak Ridge was still favoured to house the device and the scientists at Princeton were worrying about their jobs, but less than a year later, in July 1974, Princeton was actually awarded the construction of TFTR and on the path of a major expansion. Hirsch went for the more conservative Princeton proposal, which used, for instance, ordinary magnets where Oak Ridge wanted superconducting magnets. In the end the decisive factor was the projected cost of the device, which for Oak Ridge was three times what Princeton was proposing.

In this period each of the four major fusion laboratories (Oak Ridge, Princeton, Los Alamos and Livermore) lived with the constant threat of their main-line experiments being terminated. Each of them tried to find a secure niche that would be essential to the overall programme progress. Oak Ridge was historically the most vulnerable of the four and hence it zealously sought the role of host for the tokamak feasibility experiment. This posed the danger for Princeton of ceding its pre-eminent position in fusion research to Oak Ridge and forced it to submit its own proposal against its own prior best judgement. It accepted a D-T experiment which it actually did not want, but was not in a position to refuse either. This episode shows how political and scientific forces interact in such a government sponsored programme and that in the competition for resources scientific judgement often has to take a backseat to political and pecuniary considerations.

The great danger in all this was that a failure of TFTR might jeopardize the entire fusion programme. But when would it have failed? What were its objectives? Its 1976 Technical Requirements Document described the objectives of TFTR as: (1) demonstrate fusion energy production from the burning, on a pulsed basis, of deuterium and tritium in a magnetically confined toroidal plasma system; (2) study the plasma physics of large tokamaks; and (3) gain experience in the solution of engineering problems associated with large fusion systems that approach the size of planned experimental reactors. The first objective was to be satisfied by "the production of 1–10 megajoule (MJ)[1] of thermonuclear energy (per pulse) in a deuterium–tritium tokamak with neutral-beam injection under plasma conditions approximating those of an experimental fusion power reactor". A plasma temperature of 5–10 keV, a density approaching 10^{20} particles per cubic metre, and an energy confinement time of 0.1 s would be required (roughly a factor 100 below the Lawson criterion).

In July 1974 the price tag for the experiment stood at $228 million dollars, soon going up to $300 million, and at the end of the construction period in 1982 it had ballooned well beyond that. The eventual costs were almost $1 billion. One reason for this was that, due to the positive results with PLT, the objectives were amended in 1978 and 1979 and additional funding added to reach the goal of **scientific breakeven** for the D-T plasma.

Breakeven

At this point it may be useful to dwell in some detail on the term 'breakeven'. It is a word that has a seemingly simple meaning, one would think: breakeven has been achieved when the power released by the nuclear fusion reactions is equal to the power used to get the fusion reactions going, in other words

---

[1] 1 MJ is not a great deal of energy; a 100-W lightbulb will burn it away in less than 3 h.

when what goes in is equal to what comes out. With some creativity in the choice of adjectives and by being misleading about what is meant by "what goes in" or where it goes in, the fusion community has managed to make it a very confusing term.

We first define the **fusion energy gain factor** (expressed by the symbol Q) as the ratio of the fusion power produced in a nuclear fusion device to the power required to keep the plasma heated (P_{fus}/P_{heat}). This latter power, P_{heat}, should not be taken to mean all the power used to keep the reactor going or to bring the plasma into the state in which fusion reactions can start to occur. In fusion parlance, it is only a fraction of that power, namely just that part of the power that must actually be injected into the plasma to keep it at the required temperature, i.e., the external heat still needed while fusion reactions are actually going on so long as the helium-4 nuclei produced are insufficient to keep the plasma at the right temperature. And only that part of the external heating that actually goes directly into the plasma, so not including the power that is lost before such heat enters the plasma. All other losses and power used for the magnets, cooling, etc., are not included either. As we will see, this P_{heat} is actually only a tiny amount of the total power used and it is amazing that a term like 'breakeven' can be used in the way it is here. When really talking about breakeven, one should of course include all the power used in running the reactor itself and powering it up, including all the losses suffered in the process. This is not done here.

With this definition of Q, until at least $Q = 5$, self-heating of the plasma (by the helium-4 nuclei from the fusion reactions) is not sufficient to make up for inevitable losses, such as neutrons that fly off unimpeded by the magnetic field. External heating still remains necessary.

If Q increases past this point, increasing self-heating eventually removes the need for external heating. The reaction then becomes self-sustaining, a condition called ignition, and no further external energy input is needed. From the definition of Q it follows that this corresponds to Q being infinite as P_{heat} is then zero. This is generally regarded as highly desirable for a practical reactor design. It does not mean that such a device produces an infinite amount of energy or even a net amount of energy, far from it, only that the fusion reactions continue for some (short) time without requiring external heating of the plasma. The objective for TFTR and JET was $Q = 1$ and ITER is designed to give a Q value of at least 10.

For fusion systems several types of breakeven have been defined, the first being **scientific breakeven**. It is normally (and confusingly) just called breakeven and corresponds to $Q = 1$. This is what fusion devices have been aiming to achieve for the past 40 years. It is the situation in which the power released by the fusion reactions is equal to the power required to heat the plasma (a system at $Q = 1$ will still cool without external heating).

Next we have **engineering breakeven**, which is when sufficient power can be generated from the fusion power output to feed back into the heating system to keep the reactor going (recirculation) (up to $Q = 5$).

Economic breakeven applies to a machine that can sell enough electricity to cover its operating costs, while **commercial breakeven** is the situation when sufficient power can be converted into electric power to cover the costs of the power plant at economically competitive rates and any net energy left over is enough to finance the construction of the reactor. This is eventually the only breakeven that matters; a power plant that cannot achieve this will not be economically viable.

Finally, we have the peculiar concept of **extrapolated breakeven**, which is when (scientific) breakeven is projected for a reactor hypothetically using deuterium-tritium fuel from experimental results using only deuterium as fuel, by scaling the reaction rates for the two fuels. It would be better named wishful breakeven, as it has no basis in fact.

Within this framework it is disturbing, but at the same time rather sad, to read the APS News of August/September 1995 (volume 4, no 8) in which the Chair and Vice-Chair of the APS Plasma Division lambaste US Congressman Rohrabacher for stating that there is no and will not be any breakeven in fusion for many years to come. They call his statement "fantastically incorrect", and continue by saying that "[i]n fact, breakeven conditions have already been achieved in the JET tokamak operating with deuterium plasmas. In 1996, when tritium will be introduced into JET, it is expected to generate more fusion power than is put into the plasma." Here they invoke extrapolated breakeven. From the results obtained on JET with purely deuterium plasma they confidently state that breakeven ($Q = 1$) will be a certainty when the plasma is changed to deuterium–tritium. An objectionable statement by the APS Plasma Division. Very close to a straight lie and in any case "fantastically incorrect", and nobody in the science community has taken the trouble to point this out or correct it. In his statement Rohrabacher probably meant *commercial* breakeven as there is no reason for him (and most people) to be interested in any other of the more dressed-down forms of breakeven. The concept of breakeven as used by the APS includes only a fraction of the external power used. But even that kind of breakeven was not achieved by JET. In 1997, as we will see below, in its tritium operation the JET tokamak achieved $Q = (16\ \text{MW})/(24\ \text{MW}) \approx 0.67$, meaning that just 67% of the plasma power input came out as heat. That is the best that has ever been achieved. The TFTR result was even worse. So when the same publication says: "The fact is that the fusion community no longer considers breakeven an important scientific challenge: several years ago it began looking beyond breakeven at fusion self-sustainment (ignition) and issues that affect fusion power plant size, cost and complexity," they are again not telling the truth, but just complete bullshit. None of that is even close to being realised today, let alone in 1995. Apparently Rohrabacher understood more about fusion than these self-proclaimed experts, who, as a 2015 study has shown, are more likely to fall for made-up facts. Is it really necessary, one wonders, for scientists to stoop so low to keep their fusion fantasy alive?

TFTR Continued

Design studies for the TFTR began at PPPL in January 1974 before the start-up of PLT. The dimensions chosen were about a factor of two larger than those of PLT. As can be seen in the cutaway model of Fig. 9.3 (and in Fig. 9.2), the cross-section of the vessel is circular, contrary to many other tokamaks built around the same time and later, like JET and JT-60U, which had D-shaped cross-sections. PPPL stuck to the tried and tested circular design to keep their device as simple as possible. Simplicity is not always a good idea and here it was undoubtedly a mistake, as it would make it difficult to achieve high-confinement mode and/or to install a divertor. TFTR had to do with a limiter.

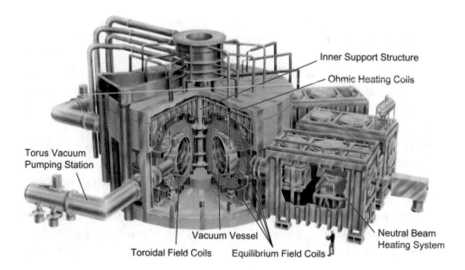

Fig. 9.3 Cutaway model of the TFTR, showing the magnetic field coils, the vacuum vessel, and the neutral-beam heating equipment

Heating was a major issue on tokamaks of this size, as ohmic heating alone was no longer sufficient and becomes negligible at the temperatures to be achieved. They all start operation without any additional heating, which is then added later and upgraded over the years. For the TFTR, heating by neutral-beam injection (NBI) and by ion cyclotron resonance heating (ICRH) was used (see Chap. 5), and demonstrated to be reliable.

With a major radius of $R = 2.65$ m, TFTR was considerably larger than any tokamak built so far. Its minor radius was 1.1 m and the plasma volume about 38 m^3.

Construction started in April 1976 and first (hydrogen) plasma was produced in December 1982, nearly nine years after the conceptual design study in 1974, showing how the construction time increases with each new generation of fusion device. This was followed by deuterium plasmas in the mid-1980s. Experiments with D-T plasmas (50% tritium and 50% deuterium) were planned to start a few years later, but were eventually delayed until 1993. In the end TFTR was scooped by the European JET device at Culham in the UK which used a D-T mixture for the first time in November 1991, albeit with only 10% tritium. TFTR and JET are still the only tokamaks in the world that have worked with a real D-T fuel mix. As said, tritium has been avoided because of its radioactivity. So, here we have the peculiar situation that energy should be produced with a fuel everybody is reluctant to burn.

From the outset it was clear that TFTR would not reach breakeven (i.e., the very limited form of breakeven ($Q = 1$) as discussed above), but experiments went ahead anyway. Despite the complications introduced by the use of tritium, TFTR operated routinely and reliably from the start of D-T operations in November 1993 to the completion of experiments in April 1997. More than 1000 shots of D-T experiments were carried out with a peak fusion power of 10.7 MW, which was the first time that any real fusion power was produced in any device. It was presented as being enough to meet the power needs of more than 3,000 homes, although just for a few seconds, or 'transient' as they tend to call it, which of course sounds much posher. It is a sly way to present the result: the use of 3,000 makes it impressive, as 3,000 is a fairly big number, but it is actually the same as "enough to power a single home for just a few hours", which does not impress anybody. This value should be compared to the more than 30 MW of input power used to heat the plasma (i.e., just the power that went into the plasma; the neutral-beam power alone, which only partly went into the plasma, was already 39.5 MW), indicating how far short of scientific breakeven the TFTR has remained (Q was just about 1/3). The total *electric* power consumed for the production of this 10 MW of *thermal* fusion energy amounted to a staggering 950 MW, so a factor hundred would still be needed to approach any useful sort of breakeven.

TFTR has so far been the only device that has actually attempted to reach breakeven. To reach breakeven, the system would have to meet several goals at the same time, a combination of temperature, density and confinement time, i.e., not only for the triple fusion product, but also for the individual components in the triple product, e.g., the temperature alone had to be 200 million degrees. In spite of considerable effort, the system could at any given time only demonstrate any one of the required values. One of the problems was that the use of neutral beams for the heating caused confinement times to get worse as the plasma got hotter, and it remained in this predicament, dubbed the "low mode" until, quite by accident, a so-called "supershot regime" was discovered, a sort of H-mode. In this operating regime the number of fusion reactions among deuterium nuclei is up to 25 times higher than previously observed. The supershot regime was an experimental discovery and apparently particular to TFTR, as it has not been reproduced anywhere else. These supershots, as measured by the Lawson triple product, enhanced performance by a factor of about 20 over comparable L-mode plasmas. The value of the triple product remained about a factor of 10 short of the required value quoted in Chap. 5.

Although it is bluntly stated on one of the PPPL web pages that TFTR accomplished all its research goals (both its physics and hardware goals), breakeven ($Q = 1$), which definitely was one of its goals, and arguably the most important one, was never achieved.

Overall 99 g of tritium was processed. TFTR set a world record plasma temperature of 510 million degrees (about 30 times the temperature in the middle of the Sun). Such records are announced with great fanfare, but are not really such a great deal; and it remains unclear what the significance of such a record temperature actually is, as it is well beyond the 200 million degrees or so required for (commercial) fusion. TFTR remained in use until 1997 and was dismantled in September 2002, after 15 years of operation. Its shutdown was due to Congressional budget cuts in the US Fusion Research budget.

There can be no doubt that from a physics point of view TFTR was a successful experiment, but it was also clear to everybody that, in spite of Hirsch's pronouncements, commercial nuclear fusion power was still very far from being in sight. No amount of funding could have remedied that. Every experiment teaches you something, solves some problem or perhaps even several problems, but the question is whether TFTR solved any of the problems that needed solving for bringing controlled fusion closer to its real-isation. One final point to note is that the pulse length for TFTR was 2 s at the most, so the total deuterium–tritium time was something like 2000s or half an hour, and that for 1 billion dollars.

Joint European Torus (JET)

First discussions on constructing a big tokamak in Europe started in the summer of 1970 and culminated in the recommendation that such a device should be built. Europe, including Britain from 1973 when it joined the European Communities, set its sights on a machine that would go beyond 1 MA for the plasma current. The cost of such a machine would be beyond the means of any individual country and a European collaboration was proposed.

The objective of the Joint European Torus (JET), as it was called, was held somewhat vague by stating that its main goal was to obtain and study plasma in conditions and with dimensions approaching those needed in a fusion reactor. There was no mention of net production of fusion power or of reaching breakeven or ignition, nor how close to reactor conditions the results had to be, although some fusion reactions in a D-T plasma were envisaged. It was all sufficiently vague for it to be declared a success whatever the outcome would be. From the 1976 design report it is clear though that the aim was breakeven.

The vagueness of JET's objectives is not all that surprising as very little was actually known about the confinement properties of the extremely hot plasmas required and it was necessary to discover the rules governing confinement in plasmas closer to reactor conditions.

One of the key decisions in the design was the geometry. Should the plasma vessel be circular or elongated, and what should be the aspect ratio, R/a? For the toroidal field coils a D-shaped profile was chosen as in that configuration the coils experience the lowest mechanical stress, exerted by the confining magnetic fields which push the coils towards the central column of the tokamak. The coils were allowed to be moulded by the magnetic forces, to find their own equilibrium. This naturally resulted in D-shaped coils and consequently a D-shaped vacuum vessel inside them, 60% taller than it was wide. It implied that both a circular and a non-circular (elongated) plasma were possible. From the calculations by Artsimovich and Shafranov it was known that a D-shaped plasma might give better performance as it would give a higher *beta*. There was little proof of this yet, but it was believed that a D-shape would allow JET to go beyond 3 MA for the plasma current.

The plasma current was considered to be the main element defining plasma performance and α-particle confinement. Therefore, JET was designed to have a D-shaped plasma cross-section, which allowed, at the chosen elongation ($b/a = 1.7$), a 1.7 fold increase in plasma current over that of a circular plasma. If this choice proved to be wrong, the plasma could still be forced to be circular and the initially planned plasma current could still be achieved.

Apart from a bigger volume and increased current, a more important aspect of the elongated shape, although not yet known in the design stage, was that it would later allow for the installation of a divertor.

The aspect ratio R/a was decided on the basis of minimising costs, and the optimum value was found to be between 2 and 3. The value chosen was 2.4. With this aspect ratio and elongation, the conveniently rounded dimensions $R = 3$ m and $b = 2$ m give a satisfactory minor radius of $a = 1.25$ m. These were the dimensions chosen for the JET design. They would make the volume of JET more than 100 m^3, almost 3 times the TFTR volume. JET implied a gigantic step forwards, at any rate in size.

The next problem that had to be addressed was heating. Just as in TFTR, ohmic heating in JET would be negligible. The question of how much heating was needed to reach interesting temperatures could not be answered. The fundamental uncertainty in this respect was energy confinement. How the confinement time depends on the size of the plasma and on the magnitude of the current and applied magnetic field was unknown. A combination of neutral-beam injection, ion cyclotron heating and lower hybrid heating was eventually used to heat the plasma.

A further problem was the stability of the plasma. The theories then available were hardly relevant to real tokamak plasmas, so studies were commissioned to try to improve the situation. This resulted in the development of the first numerical code for calculating stability in a fully toroidal geometry, making it possible to explore the stability of any proposed JET plasma. One of the instabilities was due to the choice of a vertically elongated plasma for JET, which implied that the vertical drift was not completely cancelled. Without stabilisation this would lead to an extremely fast instability in the vertical direction. However, the simulations demonstrated that the drift rate was slow enough for it to be counteracted by using additional magnets and an electronic feedback system.

Politics

The proposal for the design was presented in September 1975. In the mean-time the political bickering about the site of the experiment had already started. It would bring the whole project to the brink of disaster. JET was a high-profile international project and many EU nations were keen to have it on their soil, with initially five countries vying for the prize. By favouring sites with fusion expertise as compared to nuclear expertise, the competition was whittled down to Culham in the UK and Garching in Germany, but further progress seemed impossible. For two years European leaders were locked in debate, during which time half of the design team left, moved back to their native countries or accepted jobs elsewhere and the JET project came within a hair's breath of its demise, and this while no one had argued against building JET. Everyone wanted it to go ahead, but only after the pie had been sliced properly and to everyone's satisfaction. British prime-minister James Callaghan got personally involved and started to lobby German chancellor Helmut Schmidt, French president Valéry Giscard D'Estaing and everybody else who had a say in the matter. He even threatened that, if the European Community failed to build JET, the UK would make its own arrangements and build it at Culham anyway, perhaps with help from other countries (Iran, for instance, showed an interest). It did not help; no decision was taken in 1976 and most of 1977.

Then help, although uncalled for, came from an unexpected quarter. In October 1977 a Lufthansa plane was hijacked over the Mediterranean by Palestinian terrorists. The objective of the hijacking was to secure the release of imprisoned leaders of the German Red Army Faction (also known as the Baader-Meinhof Group). After several intermediate stops the plane ended up in Mogadishu, Somalia. Five days of negotiations with the terrorists followed after which the plane was stormed by German commandos, who shot the four terrorists and freed all passengers unharmed. The German team had been accompanied by two members of the British Special Air Services (SAS) who blinded the terrorists for six seconds with special magnesium-flash grenades, allowing the Germans to successfully storm the plane. So, when Callaghan visited Schmidt one day after the end of the hijacking, he was greeted emotionally and thanked profusely by Schmidt. In the generous atmosphere of the meetings that followed the two leaders were able to resolve some of their nations' differences over European Community matters, with Schmidt acquiescing to the siting of JET at Culham. Four years after the design team had started its work, of which the last two years had been spent in a frightful deadlock, the issue was finally settled. The delay had put them firmly in second place to TFTR, whose ground-breaking ceremony had been held in October 1977, a few days after the end of the hijack of the Lufthansa plane.

Construction could start. Participation in the project had by now been extended to 11 nations with Sweden and Switzerland also becoming partners, while Greece joined in 1983. Figure 9.4 presents a cutaway view of the tokamak (without the heating and diagnostic systems).

The figure shows the layout of the device with its major components. The innermost element is the vacuum vessel, which holds a vacuum in which the pressure is less than one millionth of atmospheric pressure. The magnetic field coils for producing the toroidal magnetic field consist of 32 D-shaped coils enclosing the vacuum vessel. The coils are to carry currents for several tens of seconds, and consequently have to be cooled, for which water is used. The magnetic field exerts a force of up to 600 tons on each coil, acting on the material in opposite directions and tending to stretch it; this is to be borne by the tensile strength of the copper. The total force on each coil is almost 2,000 tons, directed towards the major axis of the torus. A further force arises from the interaction of the currents in the coils with the poloidal magnetic field. The current in the toroidal field coils crosses the vertical component of the poloidal field in opposite directions in the upper and lower halves.

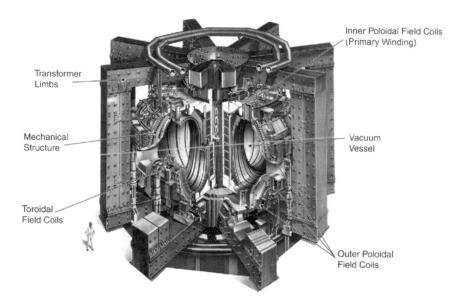

Fig. 9.4 Cutaway view of the JET tokamak. The heating systems and the many diagnostic systems are not shown in this view

This produces a twisting force which, in the JET design, is borne by an outer mechanical structure.

The poloidal field coils are horizontal circular coils, and therefore placed outside the toroidal field coils. The main poloidal field coil is the inner coil (inner poloidal field coils in Fig. 9.4) wound around the central solenoid of the transformer, to act as the primary of the transformer (see Fig. 7.2).

The massive structure of the laminated iron transformer core, weighing 2,600 tons, dominates the appearance of JET with its 8 limbs (two of which have been cut away in Fig. 9.4), enveloping the other components. The overall dimensions of the device are about 15 m in diameter and 12 m in height.

JET was designed to allow a pulse repetition rate of once every 15 min, each pulse requiring a total power of up to 800 MW—the output of a medium sized power station.

The device is equipped with extensive diagnostic systems to record plasma parameters and other data when the device is in operation. This data is required to control the plasma and the auxiliary systems, and to diagnose plasma behaviour. Diagnosis of the plasma itself involved for instance measuring the emission of varies forms of radiation. Neutron detectors were installed to measure the fusion reaction rate.

The construction of JET was completed on time, providing the first reward for the efforts begun ten years earlier. The construction was carried out without any overrun of the original budget and completed in the time of 5 years as foreseen, a major achievement. The total construction cost amounted to $438 million in 2014 US dollars (so considerably less than TFTR whose bill eventually reached 1 billion (more expensive) dollars).

Experimental Results and Upgrade

JET's experimental programme started mid-1983, when plasma was successfully produced at the first attempt. In the first few years most experiments were carried out first with hydrogen, then with deuterium, with both nearly circular ($b/a \approx 1.2$) and elongated plasmas ($b/a = 1.7$), to test the machine and optimize the plasma. Disruptions were suffered (in total 2309 over the last decade of JET operations); instabilities occurred that had to be stabilized. Disruptions in JET dropped remarkably in later experimental campaigns to only 3.4% in 2007, from a high of 27% in 1992.

Pulse lengths of up to 20 s were achieved and confinement times of up to 0.8 s, with the highest values for deuterium plasmas, more than twice the values obtained with TFTR, and found to increase with density. High electron temperatures up to 5 keV and ion temperatures up to 3 keV were reached. In general JET behaved in a similar way to smaller tokamaks. As expected, the size advantage was seen in the record energy confinement time, which was, however, still less than a second.

As discussed in Chap. 8, at other tokamaks it was discovered that confinement times went down when the heating with neutral-beam injection was increased and the plasma became hotter, the opposite of what had been expected. A solution had been found at the ASDEX tokamak with the high-confinement mode. When increasing the heating slightly, the plasma spontaneously jumped from the lower confinement of L-mode into the improved confinement of H-mode. The H-mode seemed to depend on the existence of a separatrix (last closed flux surface), which separates closed magnetic surfaces from open surfaces. JET did not have a separatrix, but the question was whether it could still achieve H-mode confinement by adjusting the currents in the external control coils and making the required change in the magnetic geometry. It turned out that it was possible (thanks to its D-shaped plasma), but only just, to create a modified geometry with a separatrix and obtain an H-mode plasma. In 1988, halfway through its experimental program, during such H-mode operation a value for the fusion triple product was achieved that was only a factor of 10 lower than the Lawson criterion, quoted in Chap. 5, at temperatures exceeding 5 keV and a plasma current up to 7 MA (for 2 s). The data showed (by extrapolation, as no tritium had yet been used) that a plasma current of around 30 MA would be required for ignition of a deuterium-tritium plasma. With lower values of the current, the risk of not achieving ignition would be great. A high current value of 30 MA would be possible, but extremely costly. A new device would have to be built, and according to a very preliminary estimate, it would cost more than ten times as much as JET. So, it was already clear that JET would not reach ignition either. The amazing thing here is that just halfway through the experimental programme of a brand new device, which had been operating for only a few years, and even before a deuterium-tritium operation had been started, a new device, which still would be just a further experiment, hence not a real reactor, was already being contemplated because JET would not be able to get close enough to reactor conditions. In this sense it was already a failure before it had even got properly started.

Both TFTR and JET had originally been designed only with limiters, although the possibility of a divertor in JET had been left open in the design report, not for achieving H-mode of course, but as a means of impurity control. Contrary to TFTR, JET did allow for the installation of a divertor without any major modifications and this was soon done.

But still before this installation in November 1991, the world's first deuterium-tritium fusion experiment was carried out with JET, not yet with 50–50 deuterium–tritium, but one pulse with 10% tritium and one with 20%, using just five milligrams of tritium, introduced into the torus by neutral-beam injection. The restriction to only two pulses had to do with the limitation of nuclear activation, i.e., inducing radioactivity in the vacuum vessel. It produced a peak fusion power of 1.7 MW and released 2 MJ of fusion energy. It ended after about 1 s with a bang coinciding with a strong influx of carbon impurities, dubbed 'carbon bloom', originating from the carbon wall of the vacuum vessel.

In the 1992/1993 period, a divertor was installed. It marked the beginning of an extensive series of divertor tokamak studies. The control of the carbon bloom and the improvement of H-mode performance were among the notable achievements.

From July to November 1997, JET conducted a tritium campaign, including three months of experiments with 50–50 deuterium-tritium fuel mixtures. In total 35 g of tritium were used. The improved plasmas with the new divertor resulted in the production of 16 MW of peak fusion power, considerably more than achieved by TFTR. The amount of power that had actually been injected for heating the plasma was 24 MW, which implied a Q value of 0.67, the highest achieved so far, but still a considerable distance from scientific breakeven. In total JET used 700 million watts of *electricity* to produce the shot that resulted in fusion particles with 16 million watts of *thermal* power, which means that JET lost a staggering 98% of the power it consumed. Only 24 million watts of these 700 million actually went into the plasma, and of this power only 16 million watts came out as fusion power. If this were converted into electricity, a further considerable part of it, probably 2/3, would be lost. Looked at in this way the 700 million watts of electricity used by JET correspond to a thermal power of about 2100 million watts, which makes the gain of 16 million watts in thermal power completely insignificant. JET was and is an enormous squanderer of energy, as is also apparent from the fact that when it is running plasma shots (typically two per hour) it draws up to 8% of the electricity supplied by the entire UK national grid.

Tritium is very expensive. Both TFTR and JET relied on what was available on the market, and it was used very sparingly in these machines. ITER, too, will rely on the global supply of tritium, but no sufficient external source of tritium exists for fusion energy development beyond ITER. In a real reactor, if it ever gets to that stage, tritium will have to be bred by the reactor itself. This concept of 'breeding' tritium during the fusion reaction is of vital importance for the future needs of a large-scale fusion power plant. ITER must lay the basis for this and we will discuss it in the next chapter.

Conclusion on JET

JET can rightly be seen as the precursor of ITER, a test-bed for ITER, and more than 30 years after its construction it is still in operation. In many ways the designs for ITER and JET are very similar. JET is the physics model for ITER and, without the results achieved at JET, ITER would not have been possible. Specific work for ITER has been and still is a prominent part of the JET program. The JET apparatus itself is still in excellent condition and, with its size and D-T capability, remains the most powerful fusion device in the world. Now that TFTR has shut down early, JET remains the only tokamak in the world able to work with D-T mixtures. It has a D-shaped plasma with a separatrix and divertor, but its shape can be widely varied. In this configuration, it is possible to reach a 6 MA current that can be maintained for over 10 s. These currents suffice to obtain energy confinement times of about a second or even longer. It is designed to work with tritium and intense fluxes of 14 MeV neutrons, and from the very beginning JET has been equipped with heavy remote handling facilities, like a nuclear machine, which is extremely important for its maintenance.

JET has not run with D-T mixtures since the ground-breaking campaign in late 1997. There nevertheless remains the legacy of this campaign in the form of more than a gram of tritium residing in the vessel surfaces, mainly on the inboard side of the divertor structure. This residual tritium presence has dictated that strict procedures of operation and maintenance must always be followed. In 2021 JET will be re-introducing tritium into its vacuum vessel for the first time since 1997, apart from some experiments with small amounts of tritium in 2003. The main purpose of the new set of experiments is to test the ITER-like tungsten wall of the vacuum vessel. JET was previously equipped with a carbon wall.

The British plans to leave the European Union (Brexit) threatened to throw the plans for JET in doubt, as the UK also plans to leave Euratom. In 2019 the UK Government and European Commission signed a contract extension for JET guaranteeing JET operations until the end of 2020 regardless of the Brexit situation. After Brexit this contract has been extended until 2024, which enables JET to support ITER in the run-up to its launch in 2025.

JT-60

About JT-60 we can be much briefer, as it was not designed to use tritium. Due to its radioactivity, tritium is a sensitive issue in Japan. In Japan no experiments with tritium are carried out, or will be carried out in the near future. The Japanese fully rely on ITER in this respect. A rather strange attitude if you want to generate a sizable part of your power needs from nuclear fusion within a couple of decades.

JT-60, with JT standing for Jaeri Tokamak (although according to others it stands for the rather unimaginative Japan Tokamak or Japan Torus), is the culmination of a long list of Japanese tokamaks constructed since the early 1970s. It is located at JAERI (Japan Atomic Energy Research Institute) in Naka, about 120 km north of Tokyo.

JT-60, with its successors JT-60U and JT-60SA, is the showpiece of the Japanese fusion effort so far. The first plans to build a large tokamak in Japan with the intention of reaching conditions in deuterium plasmas that would be equivalent to (extrapolated) breakeven in deuterium-tritium plasmas date from 1975. Here we encounter an example of the rather odd concept of extrapolated breakeven, discussed above. The rationale of working only with deuterium plasma is that all aspects of plasma physics, except those involving the fusion event itself, can be investigated in such plasmas without having to work with the radioactive tritium and without producing a high level of induced radioactivity by the high-energy neutrons released in the fusion reactions in the material of the facility, especially the vacuum vessel.

After a seven-year construction period, first plasma was obtained in April 1985. JT-60 only used hydrogen for its plasmas, so no deuterium yet. It was the only one of the three big tokamaks which had been designed from the outset with a divertor, situated inside the vacuum vessel. It proved effective in controlling impurities, but was not so successful in accessing H-mode. Divertors at the top or bottom of a torus seem to be preferable for achieving H-mode. This was corrected two years later in a first upgrade, whereby a new divertor was installed under the vacuum vessel for H-mode studies. The maximum pulse length in JT-60 is 10 s. Long pulse length is one of the characteristic features of JT-60 and its successors.

Like the other two big tokamaks, JT-60 started operation without any additional heating, which was then added in stages and upgraded over a number of years. It was equipped with NBI, ICRH and LH heating systems. When the auxiliary NBI heating system came on line in 1986, plasma confinement showed typical L-mode behaviour, with the energy confinement time degrading as the auxiliary heating power increased, as was happening in all tokamaks before H-mode was discovered. They were all inventing the same wheel and stumbling on the same problems.

The first change carried out to JT-60 in 1988 involved the installation of a lower divertor, as mentioned above. The H-mode result obtained with this new divertor was actually disappointing, but a 'serendipitous' phenomenon resembling the TFTR supershot was found. Further changes included the installation of a high-speed pellet injector to achieve high central plasma density and performance. With a pellet injector the fuel is replenished by injecting cryogenic (frozen) fuel pellets into the chamber. However, the changes did not really improve results. JT-60 remained far behind JET in its achievements and it was soon decided to go over to a proper upgrade, which resulted in JT-60U.

JT-60U started from 1991 and used deuterium plasmas (legal requirements had prevented JT-60 from using deuterium). One of its main design features is high triangularity, which measures the D-ness of the plasma shape, i.e., how much the shape deviates from an ellipse. An ellipse has triangularity zero (see Chap. 7).

World records were set in 1996 for the ion temperature and for the triple product, which is only a factor of 2 lower than the Lawson criterion. A value of $Q = 1.25$ was obtained by extrapolating the actual result with a pure deuterium plasma to a deuterium-tritium plasma. Although of interest, it remains a theoretical value and it seems that only the Japanese are impressed by it. The peculiar thing is that it is not easy to find the actually obtained value of Q in the literature. One would expect that the papers that quote the extrapolated value would also give the value it has been extrapolated from. That is, however, not the case. Even the original papers do not seem to do this; they do not even report how much fusion power has been produced. The only thing we know is that the real Q is less than one. Moreover, the values of the various reported record quantities are not obtained simultaneously. For instance, in the regime in which the record value for the triple product was obtained the extrapolated Q value remained below 0.6. One should tread carefully here. One cannot just assume any of the values reported at face value; care and suspicion must be one's companions. There is a lot of politics involved, due to the fact that certain results have been promised to politicians, who hold the purse strings, and must be presented in such a way that they are suitably impressed, to entice them to further loosen those strings.

JT-60 and its first upgrade JT-60U operated for 23 years, from 1985 to 2008. JT-60SA, although using where possible the infrastructure of JT-60, is essentially a new machine. This final version in the JT-60 series is designed to support the operation of ITER and to investigate how best to optimise the operation of fusion power plants that are to be built after ITER. It is a joint Japanese-European research and development project (within the framework of the Broader Approach agreement between the EU and Japan). SA stands for "super advanced", to indicate that the experiment will have superconducting coils and will study advanced modes of plasma operation. It is a fully superconducting tokamak capable of confining high-temperature (100 million degrees) deuterium plasmas. JT-60SA has a large amount of power available for plasma heating from neutral beams and electron cyclotron resonance heating. It will typically operate 100 s pulses once per hour. Its most important novelty in my view is that, because of its superconducting magnets, it will be able to explore full steady-state operation. The superconducting magnets must be cooled by liquid helium (about 4 degrees above absolute zero) and properly shielded to avoid them from being heated by radiation from the fusion reactions. For this purpose they are embedded in a cryostat, a vessel that can be evacuated at room temperature. In the cryostat a vacuum will be provided around the cold magnet components to minimise thermal loads. The use of superconducting magnets differs fundamentally from ordinary copper magnets which have to be (water) cooled to prevent overheating and can only provide pulsed operation.

JT-60SA will operate with a wide range of plasma shapes (elongations and triangularities) and aspect ratios (down to about 2.5), including ITER values.

The upgraded NBI system for JT-60SA consists both of positive-ion-based NBI (P-NBI) and negative-ion-based NBI (N-NBI). The P-NBI system is modified from that of JT-60U to extend the pulse duration from 10 to 100 s. Positive-ion based NBI is the traditional method, as discussed in the section on plasma heating in Chap. 5, whereby positive deuterium ions are first accelerated, then neutralised and injected into the plasma, where they are again ionised by collisions with the plasma particles and kept in the plasma by the confining magnetic field. In the case of N-NBI, the precursor ions are negatively charged deuterium ions, which are accelerated, neutralised and injected. Negatively charged deuterium ions (whereby a neutral deuterium atom must capture an extra electron) are obviously much more difficult to manufacture than positively charged ions, for which neutral atoms only have to be stripped of their electrons. Impressive progress with negative-ion based NBI systems was made in the 1990s, e.g., on JT-60U.

The construction of JT-60SA finished in March 2020 and first plasma is expected in 2021. It will then be the biggest tokamak in the world. All in all it is a very impressive undertaking, but without D-T plasmas it will remain a limited device.

Conclusion

The TFTR story related in this chapter gives the inescapable impression that it just came too early, that it was embarked on too hastily. Before such an experiment was considered, more fundamental plasma physics problems, e.g. instabilities, should have been solved. The decommissioning of TFTR left the US without its major facility. Plans for a successor at Princeton failed to win approval, partly since the US was now involved in ITER. US involvement in magnetic fusion with conventional tokamaks is almost completely channelled into ITER. No large conventional tokamaks have since been built in the US.

The JET result of 1997 discussed above is essentially where we still stand today. In spite of the fact that JET failed to achieve reactor conditions for confinement and has remained far from breakeven, the scientists involved claim without blushing that JET has been an 'outstanding scientific success'. I always get a little suspicious when words like success have to be quali-fied by adjectives, like in this case 'scientific'. Does that mean that in other respects it was not a success or a downright failure perhaps? Also what does an outstanding success mean compared to a 'normal' success? There are hardly any 'normal' successes anymore these days, it seems.

Whether JET was a success or not, now close to 25 years later no further progress has been made; all trust in a future successful outcome of the fusion enterprise is still solely based on the JET result stated above. JT-60 and JT-60U, since they had no tritium operations, have not really added anything very significant, or it must be the development of steady-state scenarios. The only thing we have and the entire nuclear fusion energy producing edifice is based on is an hour or so of experience with short-lived deuterium-tritium plasmas, most of which did not produce any or at any rate very little power.

The fact that JET did not reach breakeven was due to a variety of effects that had not been seen in previous machines operating at lower densities and pressures (making the scaling laws rather suspicious, as they are based on the assumption that no unexpected things happen). Using JET's results, and a number of advances in plasma shaping and divertor design, a new tokamak layout emerged, sometimes known as an 'advanced tokamak'. Basic features of an advanced tokamak design are a D-shaped plasma, supercon-ducting magnets, operation in high-confinement mode, and the presence of an internal divertor (that flings the heavier elements out of the fuel towards the bottom of the reactor, where a pool of liquid lithium is used as a sort of 'ash' tray). This advanced concept forms the basis of ITER.

An advanced tokamak capable of reaching breakeven would have to be very large and very expensive. ITER fits those requirements too, as we will see in the next chapter. No further progress will be made for at least another 10 to possibly 20 years, apart from undoubtedly impressive technical advances in all kinds of fields and areas related to subsystems of the ITER tokamak. Whatever ITER brings, it will certainly be more than thirty years after the 1997 JET result before we know more, after which it will again take a similar amount of time before the next step can possibly be made. Mankind will need extraordinary stamina and patience to endure such a long and uncertain wait.

10

The International Thermonuclear Experimental Reactor

In some sense this is the most important chapter of the book as it describes the culmination of more than half a century of efforts towards controlled nuclear fusion, the mammoth project of ITER. It stands for International Thermonuclear Experimental Reactor. Since to many the word 'nuclear' is as a red flag to a bull, the apparently tainted word thermonuclear has to be avoided, and ITER (pronounced as *eater*) is now supposed to stand for the Latin word *iter* which means 'way' (in the sense of direction), i.e. the way to nuclear fusion, a way that could very well be leading us into a cul-de-sac.

It is the next, and probably the last step in the attempts to harness the 'inexhaustible' source of nuclear fusion energy. It is undoubtedly a great scientific project, a jewel of technology, as the proponents like to call it, that has resulted in admirable, albeit not always smooth cooperation and collaboration between the leading nations on Earth, making it into a globe-spanning, transnational technology project. But there is a great chance that it will finally go down into the history books as one of mankind's greatest follies, born out of sheer arrogance, a true case, if ever there was one, of misplaced confidence. Not satisfied with the daily bath of energy coming from the Sun, man wants to bring the Sun to Earth and tap its source of energy at home and at will, but it is doubtful that this vision will ever come true.

ITER's history started as early as 1978 when Japan, the US, the USSR and the European Community (EC) joined in the International Tokamak Reactor (INTOR) Workshop. Soviet scientists had taken the initiative to the Workshop and wanted to be part of an international project, since

© The Author(s), under exclusive license to Springer Nature
Switzerland AG 2021
L. J. Reinders, *Sun in a Bottle?... Pie in the Sky!*,
https://doi.org/10.1007/978-3-030-74734-3_10

around 1981–1982 the USSR government had started imposing constraints on fusion research within the Soviet system. An international project was their best bet for remaining involved in cutting-edge fusion research.

INTOR was supposed to be the next step experiment in the progression from the big tokamaks to a demonstration power plant, DEMO. The big tokamaks were, however, not yet running. Far from it. The first one, TFTR, came on line only in 1982. To start the design of a new, next generation machine before the previous one has even been built is a recurring feature in fusion land, while one would think that it would be sensible to let the experience gained from such preceding devices be a guide in any new design. The fusion community has always been in a hurry, but as the saying goes: more haste, less speed.

Nonetheless, the INTOR project was a very impressive affair. All topics of relevance for a future reactor were covered in workshops and hundreds of scientists and engineers took part in a detailed technical assessment of the status of the tokamak concept versus the requirements of an Experimental Power Reactor. At the end of Phase Zero, conducted during 1979, the workshop produced an impressive 650-page report. It was the most comprehensive assessment of the status of fusion development ever undertaken. The predictable conclusion was that it was possible to undertake the design and construction of an experimental fusion power reactor based on the tokamak.

During the later phases of the Workshop it became increasingly clear that INTOR would not proceed to the design and construction stage, as support from the governments of the participants in the Workshop was lacking. In the autumn of 1987 it was folded into the new ITER Project, publishing its final report in 1988.

INTOR came too early. It should have waited for the results of the big tokamaks. In the early 1980s, as we have seen, tokamak physics had run into difficulties with additional heating, which were partly solved by the discovery of the H-mode. Extrapolation to INTOR parameters failed to give assurance that the required plasma pressure and confinement time would be reached. It had been assumed that ignition could be achieved with plasma currents in the range of 8–10 MA, but by that time JET had already operated at 7 MA and was clearly much too small to ignite. So, after some 10 years of extensive work, involving the best and brightest in nuclear fusion on the globe, it had still only produced a reactor design with parameters values that would certainly have failed. A sobering thought.

The Birth (Pangs) of the ITER Project

Evgeny Velikhov, whom we met in Chapter 8 as the leader of the Soviet fusion effort after Artsimovich's death, claims to have made the proposal for collaboration on nuclear fusion between the West and the USSR to Mikhail Gorbachev, when the latter went on his first visit to France after having become General Secretary, in March 1985, of the Communist Party of the Soviet Union. The political rationale of this proposal was scientific cooperation between the US and the Soviet Union. A 1990 CIA report adds to this that one of the reasons for the Soviets to propose such collaboration is that they "must join an international collaboration if they are to have access to a fusion Engineering Test Reactor (…) during the next 25 years. Because of economic and manufacturing constraints, they probably are unable to construct an ETR themselves" and "the advantages to the international fusion community of including the USSR in subsequent phases of the ITER programme are political, rather than technical." The cost estimates for the next generation of fusion machines ($1 billion or more) were beyond the means of Soviet national programmes and international collaboration presented a way out of this. This underlines the essentially political nature of the project.

Velikhov's initiative was taken up at the Geneva summit meeting in November 1985 between Gorbachev and US President Ronald Reagan, where it was announced that the two countries would jointly undertake the construction of a tokamak Experimental Power Reactor as proposed by the INTOR Workshop.

This proposal by Gorbachev is generally seen as the birth of ITER, while of course without the work carried out by the INTOR Workshop such a proposal would have been impossible. The INTOR Workshop is seldom mentioned though, and does not get much credit. On the other hand, would there have been an ITER project without Gorbachev's proposal to Reagan? There probably would, but perhaps in a different form and later. Independent national design efforts for INTOR-like devices had been initiated independently in the EC and the US, while Japan wanted to get involved with the US and also had its own project, but none of these projects went well. They would have continued for some time, but the realisation would soon have dawned that without international collaboration they were bound to fail.

In Geneva it had apparently been agreed for the US and USSR to prepare for fusion cooperation specifically between their two nations. The US jeopardised the project by unilaterally changing the approach and including, much to the chagrin of the Soviets, Europe and Japan, as well as involving the IAEA, under whose auspices the development of ITER was to take place.

One year after the Geneva summit, an agreement was reached with the same former INTOR participants as parties, and design work for an International Thermonuclear Experimental Reactor (ITER) started in 1988.

The First ITER Design

ITER's overall objective was to demonstrate the scientific and technological feasibility of fusion energy for peaceful purposes. Why these peaceful purposes had to be mentioned explicitly is not clear. The objective was to be accomplished by demonstrating controlled ignition and an extended burn of a D-T plasma. The device had to test and demonstrate technologies essential for a fusion reactor, including superconducting magnets and tritium breeding, but did not have to be self-sufficient in tritium and of course would not generate electricity.

As a reminder, in magnetic confinement ignition is reached when the heating by the α-particles produced in the fusion reactions matches the energy lost from the plasma, meaning that once the fusion has been set in motion, enough energy is produced to keep the plasma at the required temperature. The nuclear fusion reaction becomes self-sustaining and the fusion reactions heat the fuel mass more rapidly than various loss mechanisms cool it. The plasma 'burns' by itself; no more external heating is needed. The condition for ignition has the same form as the Lawson criterion.

Activities started in 1988 with Conceptual Design Activities (CDA), followed by Engineering Design Activities (EDA) from 1992 to 1998. The ITER process and its predecessor, the INTOR Workshop, had thus so far already spanned two decades.

The design it came up with would be capable of full inductive operation at a plasma current of about 25 MA (meaning that the plasma current is generated completely by transformer action (induction) from the central solenoid) with a burn duration (pulse length) of about 1000 s, and a toroidal field of 6 T. The toroidal and poloidal magnetic fields would be produced by superconducting coils. A longer-term objective for ITER was to demonstrate steady-state operation, so using non-inductive current drive.

The design included a divertor and a similar D-shaped plasma as JET (but more than twice the linear dimensions). The development of a robust divertor system is one of the principal challenges of ITER. In the design the maximum possible plasma would have a major radius of about 8 m and a minor radius of about 3 m. Nominal fusion power would be about 1.5 GW, with a possible extension by a factor of about 2.

The above-mentioned plasma current and size should provide the capability for ignition.

The results were laid down in the ITER Final Design Report, which is the first comprehensive design of a fusion reactor based on well-established physics and technology. With this design the bar was set extremely high: a superconducting tokamak capable of controlled ignition and extended burn in inductive pulses with a duration of about 1000 s, aiming also to demonstrate steady-state operation using non-inductive current drive in reactor-relevant plasmas with α-particle heating power at least comparable to the externally applied power.

Construction was planned to take ten years and would be followed by a programme divided into a physics phase followed by a technology phase, each of about ten years' duration. The Final Design Report put the estimated construction costs at about $5.5 billion (1989 dollars, because of inflation 30% has to be added to this number to arrive at 1998 values). Including all peripheral costs, construction would cost approximately 750 million dollars a year and annual operating costs would amount to 400 million. This must be compared with total global spending on fusion in 1998 of around 1400 million dollars ($500 million for Europe, about the same for Japan, the US $230 million and the rest by other countries). So ITER was affordable, but would gobble up more than half of the total fusion budget. All seemed positive and everybody seemed enthusiastic, but the design nevertheless failed to receive approval, as the US government judged the cost of the construction project to be unaffordable within its funding priorities. In the original ITER agreement it was agreed that the EU and Japan would each bear one third of the cost, with the USSR and US sharing the other one third. With Russia, the successor state to the defunct Soviet Union, bankrupt, the US would have to bear this one third on its own, which it did not fancy, as it would consume its entire fusion budget and probably more.

Moreover, the final ITER design came just after the failure of TFTR. The attempt to reach breakeven had been unsuccessful and the very credibility of the tokamak concept was undermined. Even the use of the word "tokamak" was discouraged.

The Americans had lost faith in themselves, in the tokamak and in ITER. Congress, helped by negative media coverage undermining ITER's credibility, instructed that the US completely withdraw from the venture. As usual, all kinds of political reasons were in play here. The Soviet Union had collapsed and the collaboration could no longer be advertised as a means to improve

East–West relations, now that the East had shrivelled to nothing with a bankrupt Russia and China not yet in the picture. Long-term concerns about energy resources and climate change existed, but there was no sense of crisis yet in this respect.

The Revised ITER Design

After the US's withdrawal, the project could either be jettisoned or a new course set out. The three remaining parties remained positive, especially Japan, but were reluctant to increase spending. Japanese scientists had also raised doubts about the burgeoning cost of the project, which now already stood at $10 billion. With Russia desperately short of money, Europe and Japan would essentially have to shoulder the greater part of the burden. It forced them to aim for a 50% reduction in cost and a redesign. The reduction in size and cost was largely achieved by compromising on the machine's main scientific objective, ignition in a burning plasma. This hugely important objective was abandoned; ignition was no longer aimed for. Instead, the new design would aim for a burning plasma in which α-particles provide at least 50% of plasma heating. It is questionable whether such a step was really necessary, whether such an outcome is worth the effort.

The three remaining parties went back to the drawing board and came up with a considerably smaller device, initially called ITER-FEAT (ITER Fusion Energy Advanced Tokomak) and expected to cost $3–4 billion. This new design was published in 2001 and proposed a tokamak device with major and minor radii as well as plasma current roughly three-quarters of the earlier 1998 design.

Figure 10.1 presents a cutaway view of the proposed new ITER tokamak next to JET on the same scale, showing how vastly larger ITER is going to be. Its plasma volume will be 840 m^3 (to JET's 100 m^3) inside a vacuum vessel of 1400 m^3. It will weigh 23,000 tons, as much as three Eifel towers, and will be housed in a building, measuring 73 m high (60 m above ground and 13 m below). All these and other figures are proudly displayed on the ITER website. They are indeed very impressive, but also deeply disturbing, since they do not bode well for any future DEMO plant or real power reactor. If they still have to be proportionally bigger, they will become far too big and, what is more important, far too expensive for a realistic power plant option.

The expected further growth in size can be seen from Fig. 10.2, which illustrates the increase in plasma volume and major radius of successive

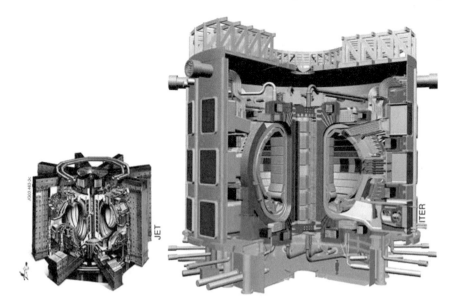

Fig. 10.1 Cutaway view of the ITER tokamak next to JET on the same scale

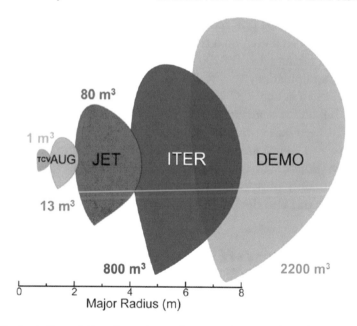

Fig. 10.2 Evolution of the dimensions of some relevant tokamaks compared with ITER and DEMO. AUG is the German ASDEX Upgrade and TCV is the Tokamak à Configuration Variable in Lausanne

tokamaks, and foresees a plasma volume of 2,200 m^3 for DEMO. The vacuum vessel will probably be about double that.

Compare this to the volume of the reactor core of a 1GW$_e$ Pressurized Water Reactor (PWR), the most common type of nuclear fission power plant, which is typically something like 35 m^3, the size of a small room, where DEMO is a massive house! For a PWR the Pressure Vessel, which contains the nuclear reactor coolant, the surrounding stainless steel cylinder (core shroud) and reactor core, is roughly 200 m^3.

After the approval of this final ITER design and motivated by the renewed prospect of a positive next step in fusion research, the US signalled in 2002 its willingness for a renewal of US participation in ITER, resulting in its rejoining the collaboration in 2003. The decision was undoubtedly influenced by the fact that China and South Korea were also negotiating their participation. The US did not want to lose out now that other countries were joining in. Its support has, however, remained lukewarm, and already in 2008 it failed to pay its financial share. China and South Korea joined the project in 2003, followed by India in 2005, indicating the broad international appeal of and support for the project.

With these additions ITER has become by far the greatest consolidation of nuclear fusion research ever, on a truly global scale, with all major economies (except Brazil) taking part. By 2005, the megaproject represented over half of the world's population, accounting for about 75% of total global GDP, estimated in 2019 at about 87 thousand billion dollars. On that scale 20, 30 or even 40 billion for a project such as ITER is just a very, very tiny amount.

The ITER Members currently taking part in the design, construction and operation of this device are the EU (actually the European Atomic Energy Community (Euratom) which comprises the 28 member states of the EU (including Britain), plus Switzerland and Ukraine), Russia, India, China, South Korea, Japan and the US.

Details of the New ITER Design

To a great extent the newly designed ITER is an enlarged version of JET. It will be a long-pulse tokamak with elongated (D-shaped) plasma and a divertor, and will produce 500 MW of fusion power (heat) in a D-T plasma with a burn length of 400 s (so the plasma will burn for 400 s, extremely long for a tokamak), with the injection of 50 MW of auxiliary heating.

The magnet system is the most important part of any tokamak. The major components of ITER's magnet system are the 18 superconducting toroidal field (TF) coils, and the 6 superconducting poloidal field (PF) coils, whose combined magnetic field will confine, shape and control the plasma inside the vacuum vessel. The coils are extremely heavy, in line with everything else in ITER. The TF coils produce a maximum magnetic field of 11.8 T around the torus. Its primary function is to confine the plasma particles. The PF magnets pinch the plasma away from the walls and contribute to maintaining its shape and stability. The poloidal field is variable and induced by both the magnets and the main plasma current. This latter current is induced by the changing current in the central solenoid, which is essentially a large transformer and the 'backbone' of the magnet system. It also contributes to the shaping of the field lines in the divertor region, and to vertical stability control. Figure 10.3 shows where the various components are situated.

The total magnet system consists of the already mentioned TF and PF coils, a central solenoid (CS) and correction coils (CC). The latter consist of 18 superconducting coils that will be distributed around the tokamak at three levels. Much thinner and lighter than ITER's massive TF and PF magnets, they will be used to control the plasma so that certain types of instabilities, such as edge-localized modes (ELMs), are mitigated. ITER's success will partly depend on its ability to control these instabilities.

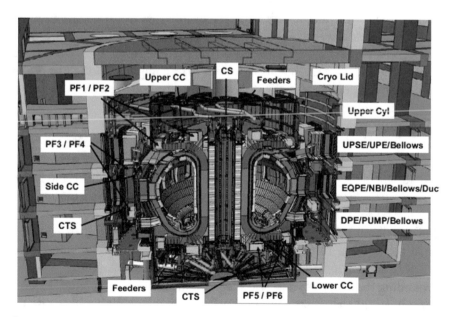

Fig. 10.3   Major components of ITER outside the vacuum vessel

The ITER magnet system will be the largest and most complex ever built. The superconducting material for both the central solenoid and the toroidal field coils is an alloy of niobium and tin (Nb_3Sn). The poloidal field coils and the correction coils use a more standard and cheaper niobium-titanium (NbTi) alloy. Both these alloys are ordinary superconductors, i.e., not high-temperature superconductors, which makes it necessary to cool the magnets by liquid helium to −269 °C, just four degrees above absolute zero.

The central solenoid consists of six separate superconducting niobium-tin coils. With its height of eighteen metres, width of four metres and weight of one thousand tonnes, it is the largest solenoid ever built for a fusion device. A maximum field of 13 T will be reached in the centre of the central solenoid, the strongest of all ITER's magnet systems. Its main function is to induce the current in the plasma, which in turn will create a poloidal magnetic field that helps to confine and heat the plasma (to an insufficient 20 million degrees).

All these superconducting coils have to be connected to power supplies. Normally, copper cables would be used for this. For ITER this is no good as the heat would leak along the copper at room temperature to the coils at −269 °C. Therefore, high-temperature superconducting (HTS) current leads will be used, transferring large currents from room-temperature power supplies to very low-temperature superconducting coils at a minimal heat load to the cooling system to keep the coils at −269 °C.

The coils of the magnet system are manufactured all over the world in Europe, Japan, Russia and China to very detailed technical specifications in order to make sure as far as possible that they will be identical. More than 1000 people worldwide are involved in the production of ITER's magnets. When in the end all these magnets turn out to be compatible and fit smoothly into their envisaged place in the system, it must be considered a great feat of engineering and management.

The entire tokamak is enclosed in a cryostat, a sort of giant thermos flask, as a thermal shield between the hot components and the cooled magnets. The toroidal and poloidal field coils lie between the vacuum vessel and this cryostat, where they are cooled and shielded from the neutrons of the fusion reactions. Imagine the enormous temperature gradient in the machine over a few metres from the 150 million degrees in the plasma vessel to the −269 °C of the liquid helium for cooling the magnets; from the highest temperature in the universe to almost the lowest possible. The cryostat, whose parts are manufactured in India, is a large (29 × 29 m), stainless steel structure surrounding the vacuum vessel and magnets, providing a super-cool, vacuum environment, as well as structural support to the tokamak. It will encase the entire reactor including all the magnets.

Many of the components of the vacuum vessel are being tested and their design adjusted and refined for ITER in smaller tokamaks around the world, like JET and JT-60.

Remote handling will have an important role to play in the ITER tokamak because of the radioactivity of the tritium to be used as a fuel and the reactor components that will be activated (made radioactive) by the fusion neutrons. When the machine is in operation, it will no longer be possible to make changes, conduct inspections, or repair any of the components in the activated areas other than by remote handling. These handling techniques must be very reliable and robust to manipulate and exchange components weighing up to 50 tons. Their reliability will also impact the length of the machine's shut-down phases.

The vacuum vessel is a hermetically-sealed steel container inside the cryostat. Both are sucked vacuum to a pressure of one millionth of normal atmospheric pressure. The magnet system is then switched on and the low-density gaseous fuel fed in. The vessel acts as a first safety containment barrier. The central solenoid will induce a current in the gas, which will be maintained during each plasma pulse, ionise the gas and transform it into a plasma. The plasma particles continuously spiral around in the vessel's doughnut-shaped chamber without touching the walls. The size of the vacuum vessel dictates the volume of the fusion plasma. The ITER vacuum vessel will be twice as large and sixteen times as heavy as the JET vacuum vessel, which has been the biggest so far. It has an internal diameter of 6 m and will measure a little over 19 m across by 11 m high. Forty-four ports (openings in the vessel wall) will provide access to the vessel for remote handling operations, diagnostic systems, heating, and vacuum systems.

The inner surface of the vacuum vessel is covered by the blanket, a thick complex structure consisting of 440 modules, also called bricks. Each brick, of which there are about 100 different types, will be about 2 m high and 1 m wide and will weigh up to 5000 kg. In ITER the blanket will serve two major purposes: (1) capture the neutrons produced by the fusion reactions and convert their energy into heat; and (2) provide shielding of the superconducting magnets from these high-energy neutrons. It is designed to withstand a thermal load of 700 MW (i.e., 700 MW in heat, the output of a small power station). In future power reactors part of the blanket will be used to breed tritium and ITER will be a testing ground for this. ITER will obtain the tritium it needs (probably the entire global supply) from the market, but tritium is expensive and the global supply, which is about 25 kg and increases by about half a kilogram per year, is insufficient to cover the needs of future power plants. There may not even be enough tritium available to start up a DEMO plant after ITER. It is therefore essential that future devices will be able to breed their own tritium. Commercial development of fusion energy will be out of the question if self-sufficiency in tritium cannot be achieved. ITER will test tritium breeding concepts by testing breeding-blanket models, called Test Blanket Modules or TBMs, which contain lithium and can produce tritium by absorbing a neutron (to be discussed in more detail in Chap. 11).

Figure 10.4 shows what the wall of the vacuum vessel will look like: the neutrons will pass through the first wall into the tritium breeding zone and from there into a coolant to extract their energy (which can be converted into electricity).

A cross-section through the torus of the tokamak in its cryostat with the various components is shown in Fig. 10.5.

The temperatures inside the ITER tokamak must reach 150 million degrees for the gas in the vacuum chamber to reach the plasma state and for fusion reactions to occur. The hot plasma must then be sustained at these extreme temperatures in a controlled way in order to extract energy. Three sources of external heating will be used to provide the input heating power of 50 MW (comparable to the amount in JET) required to bring the plasma to the temperature necessary for fusion. These are neutral-beam injection and ion and electron cyclotron heating (ICRH and ECRH).

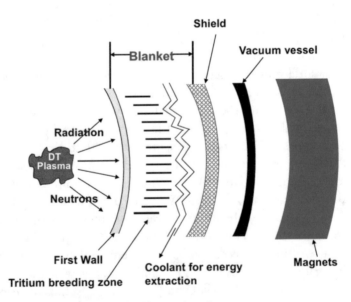

Fig. 10.4 Main layers of the tokamak vessel wall

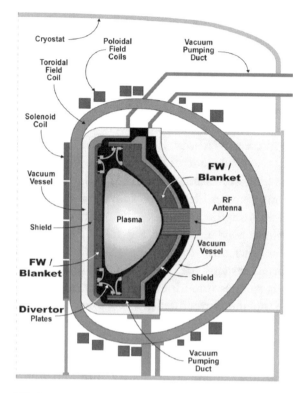

Fig. 10.5 Simplified cross-section through the torus of the tokamak, showing the most important components. The RF Antenna is used for external heating. FW is First Wall

The 600 m^2 interior surface of the vacuum vessel is covered by the blanket, which is one of the most critical and technically challenging components in ITER: together with the divertor it directly faces the hot plasma. Because of its unique physical properties (low plasma contamination, low fuel retention), beryllium has been chosen as the element to cover the first wall, the part of the blanket facing the plasma. The rest of the blanket modules will be made of high-strength copper and stainless steel. During later stages of the ITER operation, some of the blanket modules will be replaced by specialized modules to test tritium-breeding concepts.

The divertor is positioned at the bottom of the vessel at a place where the magnetic field strength is almost zero. Particles will leave the plasma flowing along magnetic field lines and then naturally fall into this 'ashtray'. In this way the divertor extracts heat (sixty per cent of the plasma exhaust is designed to go into the divertor) and 'ash' (helium) produced in the fusion reactions, minimizes plasma contamination by removing impurities, and protects the surrounding walls from thermal and neutron loads. It is one of the key components of the device, in effect a giant nuclear ashtray, covering an area of about 140 m^2. The heat flux received by the plasma-facing components of the divertor is extremely intense (10–20 MW/m^2, ten times greater than the heat load a spaceship experiences when entering the Earth's atmosphere and for much longer periods) and requires active water cooling. Only very few materials would be able to withstand the resulting temperatures of up to 3000 °C for the projected 20-year lifetime of the ITER machine. For this reason the divertor will be made of tungsten. Tungsten (with symbol W for Wolfram) has the highest melting point (3422 °C) of all elements, a low rate of erosion and thus a longer lifetime, meaning that the divertor needs to be replaced only once during the lifetime of the tokamak.

Diagnostic systems play an essential role for ITER, to make sure that the reactor operates as efficiently as possible. The device will be surrounded by about 60 measuring systems, delivering information on neutrons and on plasma parameters such as impurities and density, ion temperature, helium density, fuelling ratio and current density.

All this shows that ITER will be an extremely complex machine. In total it will have 1 million components comprising 10 million pieces that all have to work in concert to guarantee proper operation. Operators in the future ITER control room will have to play with and adjust the magnets, the external heating, the density of the gas and other parameters to stabilise the plasma against turbulence and instabilities. It may take several months, if not years, to sufficiently master and control the plasma and be able to run an experiment.

Site Selection and Construction

The New Final ITER Design Report was presented in 2001 and, after approval, construction of ITER could start, once a site had been selected. This issue turned out to be almost as difficult as designing the machine itself, in spite of the fact that there were essentially only two candidates (Japan and the EU) for siting the device.

Japan offered to host the device at Rokkasho, Aomori, in the north of its main island Honshu, and home to multiple facilities of the Japan Atomic Energy Agency. Two sites in Europe were proposed, Spain offering a site at Vandellos, about 140 kms from Barcelona, for which it was prepared to even double its contribution—upwards to $1 billion—to the project, and France the Cadarache site near Aix-en-Provence in southern France, which had long been the home of the French CEA. With the US, China and South Korea also having joined the project, there were now six parties of which Europe, China, and Russia insisted on Cadarache, while Japan, South Korea, and the US voted for Rokkasho. The US vote for Japan was seen as revenge for France's opposition to the US-led invasion of Iraq in 2003, while the support from China and Russia for the French site had clearly to do with their opposition to the Iraq war. For some time the US even backed the Spanish site to reward Spain for its support over the Iraq conflict and to spite the French. This attitude of a party that was going to enjoy a ride on the ITER carrousel while contributing a mere 10% towards its costs and had re-joined just recently was not appreciated in Europe.

It took the EU until late 2003 to decide to throw its support behind the Cadarache site. From this point on, the choice was between France and Japan and a stalemate ensued. In the competition to host the project, the main issues were global prestige and regional economic benefits. The eventual choice had little to do with engineering and instead involved 'financial, political, and social' issues. Japan was so keen to host ITER that it offered to pay a substantial fraction of the project's cost for the privilege, to which Europe responded that it would do the same. Some horse trading was going on, whereby the party that did not get ITER would host a €1 billion support facility—the International Fusion Materials Irradiation Facility (IFMIF).

Throughout 2004 negotiations continued in countless meetings, until finally, in May 2005, French President Chirac, back from a visit to Japan, declared that he had reached agreement with the Japanese. ITER would be built at Cadarache. From the details of the agreement it became clear that it involved major concessions to Japan. The EU and France would contribute half of the then estimated €12.8 billion total cost, with the other partners—Japan, China, South Korea, US and Russia—just paying 10% each. Japan would get 20% of the industrial contracts, host the IFMIF and have the right to host a subsequent demonstration fusion reactor. Finally, 20% of the project's scientists would come from Japan and it would get to choose ITER's first director-general. In November 2006, the seven Members, with India also having joined, signed the ITER agreement.

Site preparation work at Cadarache commenced in January 2007, less than a year later than originally planned in spite of all the bickering. The project almost immediately ran into delays and cost increases, resulting in a management shake-up in 2010 when the first Japanese Director-General was replaced by the second Japanese Director-General. It did not help. Poor management, especially human resources management, was partly to blame. First concrete for the buildings was poured in December 2013, but delays and cost overruns kept occurring. It is hard to keep track of the real and rumoured versions of them. Most large technological projects, and certainly any that have to do with anything nuclear, suffer delays, and some extra delays were to be expected, not only due to the complexity of the ITER device itself, but also due to the extra complexity built in by the fact that it involves a global cooperation effort unprecedented in scale, encompassing half of the world. Most delays, however, were rather mundane and due to poor planning and inadequate understanding.

In 2015 the Frenchman Bernard Bigot, former head of CEA, was appointed as ITER's Director-General and has since managed to keep the project on track. The fact is that, according to the original schedule, experiments (first plasma) were due to begin in 2016, but first plasma is now expected at the end of 2025. This will mark the end of the construction phase and the beginning of operations. But assembly will actually continue through to 2035, the planned date for the first ignition experiments using deuterium-tritium plasma. The schedule up to 2035 is shown in Fig. 10.6.

Specifications of all 1 million ITER components have been stored in an electronic package, with detailed three-dimensional models of all components. The value of the construction and manufacturing of each component was estimated and their manufacture divided among the ITER Members in accordance with the final contribution allocation key for the project (45.46% for the EU, 9.09% for the other Members). Work was started, but modifications were soon proposed which led to difficulties as components became more expensive due to the proposed changes. A multitude of Project Change Requests were submitted and had to be decided upon, resulting in delays and discord as Members refused to take responsibility for these changes or to bear the extra costs. This explains a good part of the delays mentioned above. A further complication arose as the manufacture of key systems was allocated to more than one Member. For instance, Europe, Russia and Japan are collaborating on the divertor, while it would obviously have been easier if one of them had taken sole responsibility for the complete divertor. Even strictly identical components are sometimes built in different countries, and

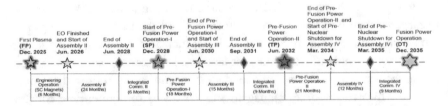

Fig. 10.6 ITER experimental schedule from first plasma

once they arrive in France they must be compatible and in full conformity with the specifications and comply with the necessary standards and requirements. All this is mentioned to illustrate how Herculean the task of constructing and coordinating the construction of ITER actually is. Nevertheless, in December 2017 50% of the construction work to first plasma was announced completed, and in 2018 it was announced that in 2021 all the main components of ITER would be on site. Subsequently, the whole gigantic jigsaw puzzle must be put together, and this will take about 1000 workers about five years. Then it will become clear whether they do indeed all fit as intended.

ITER is a machine of such daunting complexity that nobody has or could have a full grasp of it. This applies not only to the machine, but also to the equally complex organization spanning half the globe. In addition, ITER is as much a political as a technological project, with all the associated political sensitivities. From the very beginning it has been a political project, as is abundantly clear from the bickering about the ITER site. For reasons of politics and prestige France, supported by the EU, went to extraordinary lengths to make sure that Cadarache was chosen as the ITER site. The ITER Council, which oversees the project, is essentially a political body, comprised of mostly ministerial level officials. Negotiations taking place on the Council are completely different from discussions on the board of a company; the representatives will first and foremost have the interests of their respective country in mind when they commit themselves to one thing or the other. In the end their governments must be satisfied with the outcome.

We now just have to wait and see whether the latest revised schedule will be met. The assembly of the tokamak and other facilities will severely test the acumen of the ITER Organization. The Covid-19 pandemic that broke out in 2020 and stalled work in various countries for several months slowed work down and will probably result either in a further delay or in extra costs. Progress can be tracked on the ITER website where the entire construction and assembly process through to first plasma (2025) has been divided up into milestones.

When construction is finished there follows a relatively quiet period of ten years of continued machine assembly and periodic plasma operations with hydrogen and helium. These gases produce no fusion neutrons, and permit the resolution of problems and the optimization of plasma performance with minimal radiation hazards.

Cost

A big issue that tends to recur time and again in the media and is the cause of many delays is the cost of ITER. The first ITER design was estimated to cost about $5 billion. The reduced design of 2001, made necessary after the withdrawal of the US, had a price tag of less than $4 billion. Since then costs have exploded and currently a figure of more than $20 billion is quoted by the ITER Organization for the construction costs alone. Below it will be shown that total costs will be a factor of two to three higher before the game is over.

Recently, the US Department of Energy nearly tripled its cost estimate for ITER to $65 billion. The ITER Organization, however, stuck by its figure of $22 billion, which it claims is sufficient to bring ITER to 2035. The fact is though that the ultimate cost of ITER may never be known. A large part of the project is managed directly by individual Member states and the central organization has no way of knowing how much is actually being spent in these countries.

Let us try to bring some order to the ITER cost picture and get some hard figures from the ITER agreement documents and from EU documents that lay down the spending by the EU. If we know how much the EU spends we can multiply by two to get a rough idea of the total cost amount, since as host, the EU is paying about 45% of ITER's construction cost, five times the share of each of the other six partners.

The ITER Agreement mentions three types of contributions: contributions in kind (consisting of specific components, equipment, materials and other goods and services in accordance with the agreed technical specifications, and staff seconded by the Members), contributions in cash (i.e., financial contributions to the budget of the ITER Organization) and additional resources received either in cash or in kind (which are not further specified). The first type of contributions is by far the largest and each Member country has taken responsibility for a package of component parts. A nominal value has been assigned to these components and the total nominal value of the package corresponds to that Member's contribution to ITER. The Member must supply ITER with the agreed package of components and must do so at whatever the actual costs turn out to be. The components are manufactured in the Member's own country and it is perfectly possible of course that for one Member country the costs will balloon far beyond the nominal value, while others manage to remain within the original budget. In many cases they will not even be prepared to disclose how much they have spent on a certain component. The fact that each of the seven participants has its own currency further complicates matters. It is clear that such a system is very opaque and can cause problems, because in the end the real costs have to be met and, if they deviate much from the nominal value assigned to them, it could impact fusion budgets in some countries.

The ITER Agreement itself, signed in 2007, does not mention any numbers; these are contained in separate documents. They state that total construction costs for ITER will be €3.6 billion, plus an extra €400 million if required, so in total roughly €4 billion (in agreement with the figure originally quoted for the reduced 2001 design). The document also states that the figures are at January 2001 values. The EU will contribute 45.46% towards the construction costs and the other six Members 9.09% each. This only covers the construction phase, which was to take 10 years, so until 2017. For running the device, €200 million per year is envisaged and finally €800 million for deactivation and decommissioning. For these latter two phases another allocation key is used with the EU contributing 34% and the other parties 13% (US and Japan) or 10% (China, India, South Korea, Russia). One of the things we learn from this is that a distinction must be made between construction costs and running, deactivation and decommissioning costs. This is seldom done.

The next relevant document is an official EU document (EU Council Decision 2007/198/Euratom). It carries the date of 27 March 2007 and states that the indicative total resources deemed necessary for the EU part of ITER (taken care of by an organization called Fusion for Energy (F4E)) will be EUR 9,653 million (in current, i.e., 2007 values). Of this €4.1 billion would be spent between 2007 and 2016 (so for construction according to the original schedule) and €5.5 billion between 2017 and 2041 (for operation, deactivation and decommissioning). So, here we see that, compared to the ITER Agreement and accompanying documents, the cost for ITER construction alone has more than doubled (by using 2007 values compared to 2001 values, although inflation alone cannot account for such a dramatic increase in six years). Total construction costs will now amount to €4.1 billion/0.4546 = €9 billion, and total costs to about €25 billion (in the post-construction phases the EU (or better its Domestic Agency) will pay 34% of the costs, which means that its contribution of €9.6 billion − €4.1 billion = €5.5 billion must be tripled to obtain the total costs for those phases).

The conclusion is that in 2007 total ITER costs stood at €25 billion, with construction costs at €9 billion. These numbers are hard figures that follow directly from figures mentioned in EU documents and involve money that has been spent by the EU.

Let us now skip a few years and go to another EU document (EU Council Decision 2013/791/Euratom). It states that the resources deemed necessary for the EU part of ITER during the ITER construction phase for the 2007–2020 period amounted to €7.2 billion (in 2008 values). So the EU spent €7.2 billion in this first 13 year period (up from the €4.1 billion mentioned above) towards the construction of ITER. From the €7.2 billion EU contribution it follows that total construction costs in that period must have been about €16 billion. This is another hard figure: €16 billion construction costs for the period from 2007 to 2020, and the construction is not yet finished in spite of the earlier construction schedule of 10 years.

Another illuminating document is Com (2018) 445, approved early in 2019, in which spending for ITER for the six-year period from 2021–2027 is set at about €6 billion, so €1 billion per year for the EU alone for that period just for ITER construction. This, too, will have to be met by a contribution of about €7 billion from the other ITER parties, bringing the total for ITER construction up to 2027 to €29 billion (in 2008 values) (the €16 billion already spent to 2020 plus the €13 billion to be spent up to 2027). The construction phase is scheduled to end one or two years before 2027, but one can reasonably doubt that ITER construction costs will be lower than the figure just mentioned. In 2019 euro values it will be close to €36 billion (assuming an inflation rate of just 2%) or about $40 billion.

This €13 billion (in 2008 values) spent or to be spent by the EU in the period from 2007 to 2027 (the €7.2 billion up to 2017 plus the €6 billion from 2020 to 2027) is one of the few hard figures around; they all come from official EU documents and state exactly what has been spent or will be spent. If deviations occur, they have to be laid down in amended documents. The only assumption in arriving at the figure of €29 billion for ITER construction costs in 2008 values is that the contribution of the EU will be met by contributions from the other ITER parties in accordance with the ITER agreement. It concerns just the construction phase of ITER and the total value of €29 billion (in 2008 values) of total ITER construction costs spent or to be spent from 2007 to 2027 must be considered a hard figure.

Extrapolating to 2035, while assuming that from 2027 €1 billion will also be needed annually from the EU (it will probably be more), the EU alone will have spent an extra €8 billion bringing its total contribution to €21 billion (in 2008 values), about the same as the total budget the ITER Organization talks about. The period after 2027 must be considered as the post-construction phase, so if the EU spends €1 billion per year, the other parties will have to spend about €2 billion, making the total costs for that period €24 billion and total ITER costs up to 2035 €53 billion, which is not far from the US DOE figure mentioned above. And 2035 is not the end date, since experiments are scheduled to continue until 2047, after which deactivation and decommissioning still have to follow.

It may be that the figure of €3 billion per year for ITER's operating costs as assumed above is somewhat high. However, spending on ITER is currently about €2 billion per year, assuming for simplicity that the EU is paying half. Normally, when such projects get to the major operational phases the budget actually goes up. It would not therefore be unlikely that after 2025 the budget will start going up to €3 billion or €4 billion per year. Even if we assume that annual spending from 2027 does not increase and remains at €2 billion per year, the total cost will be €45 billion (in 2008 values) or $50 billion, twice the amount the ITER Organization still assumes.

Moreover, the amount mentioned here is not everything spent by the EU on ITER or in support of ITER, since it does not include the contributions from the individual EU member states towards programmes in their own countries for testing certain components or aspects of ITER. For instance, it is not clear to what extent the costs of JET, when carrying out programmes in support of ITER, have been included in the EU figure stated above.

All figures quoted at a certain time in the media or anywhere else are greatly distorted by inflation, by the use of different currencies and changing exchange rates. For a proper comparison it would be necessary to calculate them with respect to a reference year and in a single reference currency. This is virtually impossible, and makes it very difficult to get a real grip on the figures. There can be no doubt though that the costs are soaring in a fabulous, unprecedented way, which is a great worry, not so much as regards ITER, but for the prospects of any commercial electricity generation from fusion. If it costs so much to just show that power generation from nuclear fusion is in principle possible (not net power generation, mind you, but just that it is possible to have a burning plasma that can almost sustain itself for some rather short time), what will then be the cost of a real nuclear fusion power plant? And who will be able to afford it?

Conclusion

What has ITER brought so far and what can we still expect? Technical delays, labyrinthine decision-making and opaque cost estimates that have soared from five to perhaps 50 billion euro have saddled the ITER project with the reputation of being a money pit.

In fact, ITER is not really a scientific project. The underlying science has not yet come of age and is actually not yet ready for such a huge technological enterprise, embedded in global politics with the EU at the helm. Politics demands it to be a success; at stake are the prestige and credibility of the EU that has fought so hard to get the project on its soil, for essentially no other reason than the vanity of a French president. Any of the other Members can procrastinate, complain about the soaring cost or other aspects, withdraw or suspend (part of) their contribution, the EU must and will continue to foot the bill.

It has become abundantly clear by now that creating an experimental nuclear fusion reactor is an undertaking that faces both technical/scientific and political challenges. One of the general complaints of fusion scientists and administrators is that the political challenges are greater than the technical and scientific ones they are facing. That would suggest that they actually think themselves capable and able to meet the scientific and technical challenges posed. I am sure though that, if they had indeed done so, it would just have been plain sailing as far as the political challenges are concerned. The politicians would be fighting to support them. The ITER project is not the first project that promises to show that nuclear fusion can be carried out on a large scale, and apparently scientists often forget and don't understand that their confidence in getting this done is not matched by the impression outsiders get when they look in on their projects. From the early days of nuclear fusion they have greatly underestimated the scientific and technical problems, and then tended to later blame the politicians when they understandably became reluctant to foot the ever larger bills. For more than half a century they have been making promises and spending a lot of money that politicians must in the end account for. It is the first and foremost task of the scientists to convince these same uncontrollable politicians, and through them the general public, that the money will be and has been well spent. They have notably failed to do so. Where they were saying that the problems would be solved in a couple of years or in at most two decades, one now hears them say that there is still a long way ahead given "the complexity of the necessary technologies involved." Has that only dawned on them in the last few years or so, while Lawrence Lidsky for instance was saying this already in 1983?

That does not alter the fact that ITER is a beautiful and impressive machine as can be seen in Fig. 10.7. It is a wonder of technical and engineering skills, but unfortunately without the scientific (plasma physics) knowledge to match. Where the engineers have invented and developed the most amazing devices to create the circumstances in which fusion reactions

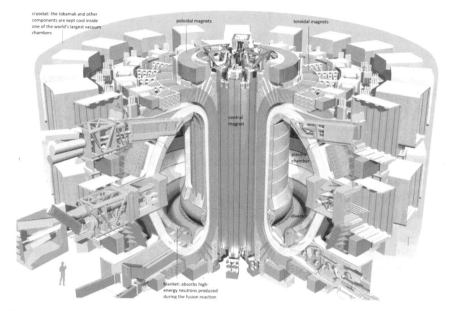

cryostat: the tokamak and other components are kept cool inside one of the world's largest vacuum chambers

poloidal magnets

toroidal magnets

central magnet

plasma chamber

divertor

blanket: absorbs high-energy neutrons produced during the fusion reaction

Fig. 10.7   Cutaway view of the ITER tokamak showing some of its vital components

can take place, basic knowledge of the behaviour of a plasma at very high temperatures in a magnetic field is still wanting.

In December 2017 ITER proudly announced that "50% of the total construction work scope through first plasma is now complete", so in ten years half of what was promised has actually been accomplished.

The first real ITER milestone will be reached around 2025 when the project hopes to produce its first plasma and "will be followed by a staged approach of additional assembly and operation in increasingly complex modes, culminating in deuterium-tritium plasma in 2035."

There is still a long way to go and all kinds of mishaps can and will occur. ITER may be hit by a violent disruption, damaging vital components of the machine (this is indeed considered a serious threat to ITER's mission and has received special attention; in 2018 the disruption mitigation system was estimated to cost €175 million, more than twice the initial estimate), old and new kinds of instabilities may play havoc with the plasma, etc.

Sometime after 2035, if all goes well and the machine behaves according to expectations, ITER will produce 500 MW of fusion power (in heat produced by fusion reactions) during pulses of 400 s and longer. This is presented as a ten-fold return on energy, which to a certain extent is true. Once the nuclear fusion reactions start, i.e., when the plasma starts to burn, the dominant mechanism for keeping the plasma at the required temperature will be heating by the α-particles produced in the fusion reactions. However, this is not sufficient; an extra 50 MW of additional heating is needed to keep the reactions going. That is the origin of the factor 10, $Q = 10$ (the 500 MW produced divided by 50 MW). If this 50 MW were not needed, Q would be infinite. The total plasma core heating needed is 150 MW, of which 100 MW must come from α-particles and 50 MW from additional heating. This 50 MW is the net heat that goes into the plasma. In order to reach the stage where the additional heating is fed into the plasma, lots of energy is needed. This energy is not included here. For the additional heating it would have been more honest to quote the 'energy at power source', so the total energy needed to get this 50 MW into the plasma. Neutral-beam heating is very efficient, but other types of heating much less so. Also completely ignored is the energy needed to bring the plasma to the required temperature in the first place, the losses encountered in the (inductive) heating process, the power needed to cool the superconducting magnets, etc. We have seen in an earlier chapter that for JET 700 MW (in electric power) was needed to produce in the end 16 MW (in heat) of nuclear fusion power. ITER is supposed to do better, in spite of its larger size, since its superconducting magnets require much less energy than JET's copper coils. The ITER website states that electricity requirements for the ITER plant and facilities will range from 110 MW up to 620 MW for peak periods of 30 s during plasma operation. Can we conclude from this that 620 MW is the equivalent of JET's 700 MW? Or is it 110 MW? Of course, ITER does not continuously draw such amounts of electric power from the grid, only during plasma operation, but also, when the device is not operating, the continuous electric power drain for its auxiliary systems varies between 75 and 110 MW.

Contrary to the ITER website, the JT-60SA website is brutally honest about what ITER will achieve: "The efficiency of the heating systems is ~40%. Other site power requirements lead to a total steady power consumption of about 200 MW during the pulse. Now the fusion power of ITER is enhanced by about 20% due to exothermic nuclear reactions in the surrounding materials. If this total thermal power were then converted to electricity at 33% (well within reach of commercial steam turbines), about 200 MW of electrical power would be generated. Thus ITER is about equivalent to a zero (net) power reactor, when the plasma is burning. Not very useful, but the minimum required for a convincing proof of principle. In ITER the conversion to electricity will not be made: the production of fusion power by the ITER experiment is too spasmodic for commercial use." Here the quote from the JT-60SA website ends and now we know, thanks to the Japanese. It can be summarised as: ITER will not establish much.

And that while ITER's website greets you with the proclamation "UNLIMITED ENERGY" in huge, boldly printed letters, while it will not produce any energy at all; it is not even intended to produce energy! Its carbon footprint will be unfathomably large. The energy used in its construction and operation will partly come from France's nuclear power stations and partly from fossil fuels, enormous amounts of fossil fuels, and will never be paid back!

We finish the discussion on ITER by summarising what the basic flaws are in the ITER project, making the *path* or *iter* it sets out unsuitable for the bright future proclaimed above:

1. **ITER is too big**. The core of an ITER based power plant would be at least 60 times more massive than a conventional nuclear fission core. And that is just the core!

2. **ITER is too complex**. The machine has roughly one million parts. Imagine the cost of doing maintenance and repair on such a machine.

3. **ITER is too expensive**. More than €53 billion, while any fusion power plant may well cost considerably more.

4. **ITER is too late**. Delays are slowly running into decades now, and whatever fusion may do in the future, it will not be able to contribute anything to combating climate change. It just comes too late.

5. **ITER is not safe and not clean**. ITER creates two safety issues: plasma disruptions and quenching, apart from radioactive waste and radioactive fuel. If disruptions accidentally happen, it will be expensive and dangerous. The second problem is quenching, when a superconducting magnet suddenly becomes a normal electromagnet – and releases its energy. ITER's coils contain the same energy as 10 tons of TNT. Such quenching has already happened 17 times in tokamaks. It causes overheating and melting of components. These topics will be discussed in Chaps. 16–18.

6. **ITER needs new technologies**. No solid material can withstand the neutron flux from an ITER-like steady state. See Chap. 11 for more on this.

11

Problems, Problems, Problems ......

ITER and its possible successors face a number of technical problems that have to be resolved before commercial energy generation can become a reality. The most important ones are finding suitable materials that can withstand the onslaught of the high-energy neutrons produced in the fusion reactions, the breeding of tritium such that self-sufficiency in tritium can be achieved, and the avoidance and/or mitigation of disruptions and instabilities (especially ELMs). In this chapter we will discuss these in some detail. Other problems related to safety and the environment will be dealt with in later chapters.

Development of Materials

Materials are required that do not easily become radioactive on exposure to very high-energy neutrons and maintain good mechanical and thermal properties under extreme conditions (considering that temperatures within a fusion reactor range from tens of millions of degrees down to a few hundred degrees below zero).

The neutrons produced in D-T fusion reactions have an energy of 14.1 MeV, much higher than the less than 2 MeV on average for neutrons emitted in nuclear fission reactions. There is no experience whatsoever of such hard neutrons. Since neutrons carry no charge, they cannot easily be stopped and are able to penetrate several metres into material before coming to a standstill. As they move within the materials, they create microscopic

© The Author(s), under exclusive license to Springer Nature Switzerland AG 2021
L. J. Reinders, *Sun in a Bottle?... Pie in the Sky!*,
https://doi.org/10.1007/978-3-030-74734-3_11

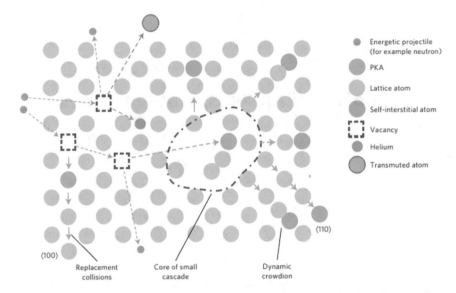

Fig. 11.1 Schematic illustration of radiation damage, as explained in the text

changes in the structure of the material resulting in degradation of physical and mechanical properties (Fig. 11.1). The first wall of the reactor vessel will be most seriously affected by this. Once the damage has accumulated, the reactor must be shut down, the vacuum vessel vented, the highly activated (radioactive) first wall replaced by remote handling, a new one installed, impurities pumped out, etc., before the reactor can be restarted. This kind of interruption in running a fusion reactor cannot be avoided and could well be more costly than the proceeds from electricity production from the reactor.

Such radiation-induced damage (neutron activation) is a huge problem for fusion materials and its study must necessarily go far beyond the damage level in fission materials. Fission materials have always been tested in experimental fission reactors, but no facility exists that offers a suitable flux with the neutron spectrum required for fusion materials research. For this we need a source that produces neutrons of roughly the same energy as generated in such fusion reactions. This is an indispensable step towards the possibly successful development of fusion. None of the currently available neutron sources is adequate for testing neutron activation of fusion materials. Materials are very sensitive to the specifics of the irradiation conditions and material tests require the neutron source to be comparable to a fusion reactor environment. This is a very serious problem and the failure to master the challenges of structural and functional materials would be a further reason for commercial electricity generation by fusion to remain but a dream for humankind.

Neutrons of a given energy spectrum and flux induce structural changes in materials. The flux is the number of particles that goes through each square metre per second. Apart from the flux there is also the fluence, which is the number of neutrons that has gone through a square metre during the whole life of the material, i.e., the flux integrated over time. In fusion power plants the neutron flux will be of the order of 10^{18} particles per square metre per second and their energy is, as mentioned, a staggering 14.1 MeV, resulting in an energy flux of about 2 MW/m^2 hitting the wall. Each neutron impact will give rise to a cascade of collisions which will displace many atoms from their original positions in the material. This has been illustrated in Fig. 11.1. The structural change or damage to the material is expressed in displacements per atom (dpa). Its definition is the number of times that an atom is displaced for a given fluence (i.e., for a given time of exposure to the flux), whereby one dpa means that, on average, every atom within the material will have been displaced once.

In elastic and inelastic collisions, a significant part of the neutron energy is transferred to the recoiling atom (primary knock-on atom (PKA)). The atoms are scattered in various directions and with various energies, are left in an excited state, and can in turn displace secondary knock-on atoms. The PKA atom loses its energy by damage production as well as by ionization. The displaced atom, called a self-interstitial atom and leaving a vacancy, can 'annihilate' with another vacancy or can share a regular lattice site with another atom, resulting in a 'crowdion' (an extra atom inserted (crowded in) within a row of atoms in the material). Significant amounts of protons and helium particles are created by transmutation reactions, leading to irradiation embrittlement. The neutrons will also generate transmutation reactions in the material (conversions into another chemical element). This damage can be as important as displacement damage. In such transmutations significant amounts of hydrogen and helium will be released and lead to a presently undetermined degradation of structural materials after a few years of operation. He-induced embrittlement, for instance, already a concern for fission materials, becomes even more critical for fusion materials.

The new tritium campaign in JET, which is expected to start in 2021, will in particular also address these issues. JET will generate neutron yields large enough to cause easily measurable activation in materials and degradation in their physical properties. JET will obtain the first complete and consistent "nuclear case" for a tokamak using D-T fuel.

Whereas in ITER structural damage in the steel of the device will not exceed 2 dpa at the end of its operational life, damage in a fusion power plant, which will have to deal with hard neutrons and tritium on a continuous basis, is expected to amount to 15 dpa per year of operation. After many dpa the material will swell or shrink and become so brittle as to be useless. Therefore, a commercial fusion reactor will require materials capable of withstanding 150 dpa. In particular, the choice of material for the reactor vessel's first wall and for the divertor is one of the toughest engineering problems faced by fusion research.

Much useful information on materials can be obtained from the experience of fission reactors and from computer simulations. For ITER such information is sufficient, but for the design and construction of any fusion reactor subsequent to ITER, a neutron source with a suitable flux and spectrum is an indispensable facility, and results must be available before construction of such a follow-up plant can start.

In Chap. 10, when discussing the squabble over where to locate ITER, we noted that, to soften the blow for not having secured ITER, Japan will host a support facility—the International Fusion Materials Irradiation Facility (IFMIF) . This facility will carry out testing and qualification of advanced materials under conditions similar to those of a future fusion power plant. It has been in the pipeline for a very long time now and, in spite of the fact that it is urgently needed, it is still not clear when construction of the facility will actually begin.

The IFMIF project was started in 1994 as an international scientific research programme, carried out by Japan, the European Union, the United States, and Russia. Since 2007, the project is being pursued by Japan and the European Union under the Broader Approach Agreement, an agreement for complementary fusion research and development between Euratom and Japan that also includes the JT-60SA. The IFMIF materials testing facility is currently in its engineering validation and design stage.

IFMIF's basic tool will be an accelerator-driven neutron source, producing a high intensity fast neutron flux with a spectrum similar to that expected at a fusion reactor's first wall. The IFMIF tool will consist of two linear accelerators, creating beams of deuterium nuclei that strike a flowing lithium target and by doing so provide an intense neutron flux of about 10^{18} neutrons m^{-2} s^{-1} with a broad peak near 14 MeV, exactly as required to look like nuclear fusion neutrons. With this neutron flux, damage rates higher than 20 dpa per year of full operation are expected to be reached in a test volume of 0.5 l. Note how small this volume actually is, just half a litre.

Of course, in spite of the fact that IFMIF has not yet got off the ground, a successor has already been thought of. It is impossible, it seems, to escape the relentless pressure of thinking ahead. In fusion it is very much the standard thing to do. DONES, which stands for DEMO Oriented Neutron Source, is a future version of IFMIF and will take over when the latter's operation, which has yet to start, has come to an end. It will be designed to mimic the conditions of neutron irradiation in DEMO, the demonstration reactor that is to follow ITER, and to allow scientists to test materials in an environment mimicking DEMO conditions and characterize candidate fusion materials. Keen to have some flows of money directed to their countries, Croatia, Poland and Spain have expressed an early interest in hosting the facility, and it has now been agreed to host it in Granada, Spain. Like IFMIF, it will be built within the framework of the Broader Approach agreement between the EU and Japan. It is not clear to me though why a completely new facility has to be built for this; one would think that IFMIF could easily be upgraded to do the job. The neutrons coming out of DEMO and ITER are the same.

The First Wall

As mentioned, one of the first and greatest challenges that must be met is to find a suitable material for the first wall, which has to bear the brunt of the neutron onslaught. In ITER the first wall consists of 440 detachable panels covering the 610 m^2 surface of the vacuum vessel. Depending on their position inside the vessel the panels are subject to different heat fluxes. There are therefore two different kinds of panels: a normal heat flux panel designed for heat fluxes up to 2 MW/m^2 and an enhanced heat flux panel that can withstand heat fluxes up to 4.7 MW/m^2. Within ITER's operational lifetime, the panels will be replaced at least once.

Figure 11.1 illustrates various processes that can take place when neutrons hit the first wall. In early tokamaks stainless steel was used for the first wall, but that is clearly not a high-temperature resistant material. Current tokamaks use carbon fibre composites (CFCs), which are light, strong and high-temperature resistant, but carbon cannot be used in fusion reactors that burn deuterium–tritium as it absorbs tritium (just forming a hydrocarbon as carbon does with other forms of hydrogen), depleting this scarce fuel and weakening the CFC. The same applies to carbon fibre-reinforced graphite (C/C). Tungsten was proposed as a material, but as a high Z (atomic number) material it has so many electrons (74 to be precise) that it cannot be completely ionized and the remaining electrons will radiate energy away, cooling the plasma.

In ITER the panels of the first wall will be made of high-strength copper and stainless steel with a beryllium coating on top. Beryllium has low Z (just 4 electrons), but a low melting point and must therefore be aggressively cooled. ITER will be the first fusion device to operate with such cooling (first tested on JET, which also used beryllium). But beryllium cannot be used in a DEMO device or a real fusion reactor. It melts too easily, and its retention of tritium and its toxicity would be problematic.

In short, the conditions that a first-wall material must meet are that it does not absorb tritium, has a low atomic number, takes high temperatures and is resistant to erosion, sputtering (the ejection of microscopic particles from the surface of the material when bombarded with highly energetic particles) and neutron damage. Various other proposals have been made, but no fully satisfactory material has yet been found. Promising materials that are still being investigated are silicon carbide (SiC) composites. The advantages of this material include low activation, high temperature capability, relatively low neutron absorption, and radiation resistance. One drawback is that at present there is no known method for manufacturing SiC in large quantities.

A materials testing facility for testing such materials under fusion equivalent conditions is therefore especially needed for the comparatively large step between ITER and DEMO. The neutron damage in DEMO is expected to be about 80 dpa (compared to <3 in ITER) and 150 dpa in a real reactor. The difference is due to the fact that ITER is just an experiment, while in a reactor the first wall should last for some 15 years of almost continuous operation before it is replaced.

The conclusion is that the solution to the problem is still wide open.

The Divertor

More important still is the material to be used for the divertor, since sixty per cent of the plasma exhaust is designed to go into the divertor, taking the major part of the heat load away from the first wall. The ITER divertor will be made solely of tungsten, mainly because it has the highest melting point of all metals, but also to minimize tritium trapping. The heat load on the divertor surfaces is huge, about 20 MW/m^2 (ten times higher than that of a spacecraft re-entering Earth's atmosphere), and cracking and melting under such high loads may damage the tungsten surface and shorten its lifetime, so cooling is essential. In ITER water cooling will be used, consuming huge volumes of water, but in DEMO or a fusion power plant helium cooling will be necessary. Tungsten in itself is a poor structural material, so alloys will be applied as a surface layer on a special type of steel, RAFM (Reduced activation ferritic/martensitic) steel. The Chinese-French tokamak WEST is equipped with a full ITER-grade tungsten divertor and will start to study these issues in 2021.

Tritium Breeding

ITER will obtain the tritium it needs from the market, but future devices are supposed to breed their own tritium as the global supply of tritium will be insufficient for their demands. The scarcity of tritium will be discussed in Chap. 17. A tritium-breeding blanket ensuring self-sufficiency in tritium is a compulsory element for DEMO, the step after ITER, as DEMO will need 300 g of tritium per day. ITER will test tritium-breeding concepts by testing pilot models of breeding blankets.

One of the tasks of the massive blanket surrounding the vacuum vessel is to produce tritium, in addition to the functions mentioned in Chap. 10. For ease of replacement the blanket is composed of modules. ITER will have 440 blanket modules covering, like the first wall, the entire inner wall of the vessel. The entire blanket must be inside the vacuum vessel and will be the first layer after the first wall (see Figs. 10.4 and 10.5). The neutrons first strike the first wall and then go into the blanket, where their energy is captured. The heat is taken to heat exchangers outside via hot gas or liquid coolants, and in future reactors will be turned into electricity. The tritium will be bred in the blanket from lithium and about 300 kg of ^{6}Li will be needed per reactor per year. Figure 11.2 gives a graphic representation of the basic reaction of the tritium-breeding process, which looks as follows:

$$^6_3\text{Li} + n \rightarrow {}^4_2\text{He}(2.05 MeV) + {}^3_1\text{H}(2.75 MeV)$$

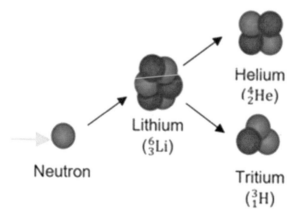

Fig. 11.2 Tritium production in the breeder blanket: a neutron strikes a lithium nucleus which disintegrates into a helium nucleus and a tritium nucleus. Neutrons are blue and protons are red in this figure

Tritium-breeding blankets will contain lithium, lead and beryllium, in addition to a structural material. The lithium can be in the form of solid pebbles or a lithium ceramic, a liquid mixture of lead and lithium, or a lithium- and beryllium-containing molten salt. In ITER six proposals for tritium-breeding solutions will be operated and tested, with different proposals coming from the EU, Japan, Korea, China and India. They are still in the design stage. The main problem is cooling and taking out all the heat for generating electricity. Blanket designs differ in the way they are cooled, with the main coolants being water, liquid metals and helium.

One of the problems with the proposed designs is that they can barely breed enough tritium to keep a D-T reactor going. The single neutron created in a fusion reaction is not enough as it can create only one tritium nucleus by hitting a lithium nucleus. Those that hit something else or nothing at all do not give rise to tritium. Therefore beryllium has been included in the blanket to act as a neutron multiplier. When a high-energy neutron strikes a beryllium nucleus, the latter breaks up into two helium nuclei and two neutrons, which can in turn react with lithium to produce tritium.

The number of tritium nuclei created in the blanket for each incoming neutron is the tritium-breeding ratio (TBR). It has so far not been possible to design a blanket with a TBR greater than 1.15, so no bigger margin than 15%. Since only a small percentage of the tritium injected into the plasma actually fuses (this percentage is called the fractional burnup) and produces a neutron that can generate a tritium nucleus in the blanket, the implication is that tritium self-sufficiency can only be achieved after several decades, which is obviously not good enough for a real reactor. The fractional burnup is only a few per cent; in ITER it is expected to be only 0.3%, so the step between ITER and a real reactor is still huge. ITER will use up most of the tritium available in the world, so there is some urgency in developing breeding blankets with higher TBR. Advances in breeder-blanket design are being made, and, as said, six will be tested in ITER, but there is no guarantee that any of the proposals will be satisfactory. The failure to develop a suitable tritium-breeding blanket will actually be a showstopper for D-T fusion, as there is simply not enough tritium around to do the job.

An added complication is that tritium is radioactive. It decays with a half-life of 12.3 years by emitting an electron (β-decay) and converting into helium-3. So, it is continually lost, 5.5% per year. The emitted electron has very low energy and cannot penetrate the skin, and even in air it can only go 6 mm, but it can be a radiation hazard when inhaled, ingested via food or water, or absorbed through the skin, so special precautions must be taken.

Disruptions

We have been going on about disruptions more than once in the preceding chapters, e.g., in Chap. 9 when discussing such events in JET and TFTR. A disruption occurs when an instability grows in the tokamak plasma with a rapid loss of the stored energy. It is a violent event that causes the current in the tokamak to terminate abruptly, resulting in the loss of temperature and confinement. The amount of heat in a large experiment like ITER will be about 400 MJ, equivalent to the explosive yield of 80–90 kg of TNT, the power of a fairly large bomb. In addition, another 400 MJ of energy is held by the poloidal magnetic field (created by the tokamak current). The toroidal magnetic field actually holds much more energy, but as long as the toroidal field coils remain undamaged, that energy will not be released in a disruption. Fortunately, not every disruption is violent and JET actually experienced a few thousand disruptions in the last decade of its operations. Such events will, however, be much more powerful in ITER and therefore extremely dangerous, capable of causing considerable damage. For ITER it is essential to have a very low rate of disruptions by the time it reaches D-T operation. Unmitigated disruptions at plasma currents above 8.4 MA may be so severe that they can only be allowed to happen once or twice in the machine's lifetime, and a disruption at ITER's planned current of 15 MA must consequently be avoided at all cost. Since the cause of disruptions is still poorly understood, prevention or mitigation is essential. If this turns out to be impossible, the entire structure of the device has to beefed up to be able to absorb all the energy that is released in a disruption. This will add considerably to the costs.

The damage caused by a disruption is threefold. First, the plasma's heat is deposited into the walls and causes them to vaporize in spots (this is called thermal quench). Even if most of the heat can be channelled into the divertor, there is no time for it to be conducted away and the tungsten and carbon in the divertor will also vaporize. Secondly, the plasma current decreases very rapidly (current quench) and causes a counter current to be driven in the conducting parts of the confining vessel. This current will exert a tremendous force on the vessel, and, if it is not sturdy enough, move or deform it. Thirdly, there are runaway electrons; electrons accelerated to high speed in the strong electric field formed as the thermal energy of the plasma gets rapidly lost that never find an ion to collide with and literally 'run away'. They can amount to 50–70% of the original tokamak current and are dangerous for the plasma facing components.

It is an extremely serious problem and as recently as May 2018 the Science and Technology Advisory Committee to the ITER Council characterized disruptions as "a serious threat to ITER's mission." No wonder that the disruption mitigation system is considered to be one of ITER's key systems to ensure its reliable and successful operation. In that context and the advanced state of ITER's construction it is a little surprising that only as late as 2018 a special Disruption Mitigation Task Force was established for designing a mitigation system. The strategy of tackling the problem had apparently first to be agreed upon before the Task Force could get off the ground. The design of such a system has proved challenging because of the complex physics involved in stopping runaway electrons. During plasma disruptions in ITER, a massive generation of such electrons is expected. The disruption mitigation system has to protect the plasma-facing components against the heat and the forces that arise during the disruption, and at the same time it must tame the runaway electrons, which could cause melting of the first wall and leaks in the water cooling circuits.

The concept chosen for the ITER disruption mitigation system is based on so-called shattered pellet injection (SPI), a technique developed at Oak Ridge National Laboratory and pioneered at General Atomics. SPI stops an abrupt termination of the plasma, i.e., a disruption, by shooting frozen deuterium-neon pellets into the plasma to bring the temperature down significantly and increase the density (which suppresses the runaway electrons). It is a safe way to dissipate plasma energy and to minimise the damage from the disruption to in-vessel components. The pellets are shattered into small pieces just before entering into a disrupting plasma. The largest pellets are larger than a wine cork, with a diameter of 28 mm. Despite this "enormous" size for a cryogenic pellet, several of them have to be fired at the same time to reach the required quantities to stop the worst case of runaway electrons in ITER. Several tokamak experiments are planned and in preparation to gain experience with this system, e.g., at JET and the Korean KSTAR.

It should be realised though that such techniques do not prevent the disruption, but only mitigate the consequences. The result is still a machine that has to be emptied and started up anew (TFTR for instance needed more than a month for recovery after big disruptions), not something you want to happen too often when running a power station. It is also clear that the collapse of the plasma occurs on such a fast time scale (the energy sprays out in a matter of 10 ms) that the pellet delivery time must be very precise and a few milliseconds difference in delivery times can be detrimental to the mitigation achieved. Figure 8.1 illustrates what happens in a disruption, with the relevant timescale.

Prevention of disruptions, which must be the eventual aim for a power generating device, is at a much less advanced stage, although some progress is being made. As often in nuclear fusion science, the basic knowledge is lacking and the search for a solution proceeds backwards. Disruptions occur when the system is pushed beyond or too close to its limits and control over the plasma is lost, in particular the density limit and the pressure or *beta* limit. The mechanism behind these limits is not fully understood. Impurities can also give rise to instabilities and disruption as they affect pressure and current density. An average for all tokamaks shows that in 13% of the pulses a disruption has been suffered, while JET achieved an average rate of 3.4% of unintentional disruptions in the 2008–2010 period. In ITER nearly disruption-free operation will be required, in addition to highly reliable mitigation of any disruptions that do occur, whereby disruption mitigation should be a rarely-used last resort.

In addition to the above, the α-particles produced in the fusion reactions can also be a cause of disruptions, a problem that cannot be studied before ignition has been achieved, so even ITER is at a loss here. As the α-particles cool down, they transfer their energy to the plasma, keeping it hot. Before this happens, they stream in the form of beams along the magnetic field lines and can excite so-called Alfvén-wave instabilities, electromagnetic waves that become so strong that they disrupt the plasma.

Suppression of Large ELMs

Edge-localized modes (ELMs) have shown up before in this book. They were described as a way for the plasma to let off steam when the pressure is building up. As we have seen (Chap. 8), the H-mode was discovered in the German ASDEX tokamak in 1982. For the resulting more stable confinement, a price had to be paid as pressure built up at the plasma's edge and a new type of instability appeared. The H-mode plasma abruptly experiences "storms" amid the calm, akin to solar flares on the Sun. These are called ELMs, instabilities at the edge of the plasma, as their name suggests, and they eject a jet of hot material. With each ELM, the surface of the vessel faces a sudden increase in the temperature by thousands of degrees and a large ELM can represent 5–10% of the total energy stored in a fusion plasma. It is the divertor and the plasma facing components, in particular, that will suffer from the onslaught of ELMs. They are very nasty instabilities and could easily spoil confinement in ITER, where thousands of ELMs are expected. Type-I (giant) ELMs can deposit enough energy to melt some of the beryllium wall, so their occurrence must be minimized by stimulating smaller Type-III ELMs.

The spherical tokamak MAST (see Chap. 13) at Culham has been using an ELM-mitigation technique called resonant magnetic perturbation, which consists in applying small magnetic fields around the tokamak to punch holes in the plasma edge and release the pressure in a measured way. This technique has been successful in curbing large ELMs on several tokamaks by generating a stream of smaller, less powerful ELMs that will not damage the tokamak. To utilize the technique in ITER, its design was changed to include additional electromagnetic coils that will attempt to control the ELMs by providing such magnetic perturbations. These coils will be very close to the inner wall, so they must resist large thermal expansions due to the high temperatures, while remaining reliable for 20 years or so. The final ITER design now includes such ELM coils, but of course practice must show whether they do actually cure the problem.

Another way of trying to calm down ELM instabilities was dubbed the "snowball-in-hell" technique. It consists of throwing a small pellet of cold fuel directly into the plasma, which produces a minor instability that, somehow, prevents the much larger ELMs. It is similar to the frozen-pellet injection discussed above for mitigating disruptions.

Several other ways to tame ELMs are being investigated, but the final and definitive solution has not yet been found.

Conclusion

In this chapter we have identified some obstacles, and potential showstoppers, on the way to a practical energy generating fusion reactor. In the first place the development of materials that can withstand the flux of high-energy neutrons originating from fusion reactions. Such materials are especially required for the first wall and divertor of the device. Far too little progress has been made in this respect and neutron sources with a fusion-relevant neutron spectrum must be built urgently for a reliable design of a DEMO plant or pilot fusion plant to be possible.

Secondly, there are the difficulties in breeding sufficient tritium to obtain self-sufficiency in tritium. Various breeding-blanket proposals will be tested in ITER. It is, however, very unlikely that tritium self-sufficiency will ever be achieved. The tritium-breeding ratio is bound to remain too low and the scarcity of some materials (beryllium, lithium) may also play a disruptive role in this respect (see Chap. 17).

In addition, there are the long-standing physics issues regarding disruptions and instabilities, in particular large ELMs. Instabilities and disruptions are part and parcel of fusion plasmas, and have to be lived with. It does not help of course that the physics of these phenomena is not very well understood either. The question is whether mitigation techniques can be developed that can sufficiently moderate and mitigate such events without shutting down the power generation of the plant.

12

Post-ITER: DEMO and Fusion Power Plants

After ITER the path to a commercial fusion reactor is still a very long one, even if ITER turns out to be the greatest success in nuclear fusion the world has ever seen. The route is fairly clear and consists of two further steps before one can start thinking about building a full-size nuclear fusion power plant, to be operated by industry.

The first step, mentioned in the previous chapter, consists of the construction of one or several large machines like IFMIF for solving the materials problems. Some believe that such a first step is not necessary and that the ITER experiments will give enough information for designing a demonstration reactor (DEMO), but in that case we shall have to wait for these ITER experiments, i.e., until after 2040, which in the current impatient climate is apparently not what anyone wants to do either. This impatience is understandable as the battle against climate change must be fought now. Any carbon-free power-generating option that cannot come online within a few decades will not be able to contribute much in that respect.

The second step will then be the construction of DEMO, a prototype reactor built to run like a real reactor but not producing full power. DEMO designs are already well underway even before the above-mentioned first step has been made and even before ITER has carried out any experiments. One of the amazing aspects of this is that all DEMO designs aim to burn deuterium and tritium, while the only experience with tritium is tiny, just an hour or so of experiments on JET and TFTR.

L. J. Reinders, *Sun in a Bottle?... Pie in the Sky!*, https://doi.org/10.1007/978-3-030-74734-3_12

ITER has turned out to be a cumbersome project and hideously expensive, so expensive that one would not think it feasible for any one country to be able to go it alone to a DEMO or power plant or want to run the associated risks of failure. Nonetheless, several of the current partners in the ITER consortium, which are the only serious players in the nuclear fusion power game, seem intent on doing so. The consortium that joined forces to build ITER is unlikely to continue into the post-ITER period. The spirit of collaboration that characterised the early activities on ITER has apparently not whetted the appetite for more of the same for subsequent stages of the fusion effort, despite the fact that future devices will probably be even more costly and more complex than ITER and might greatly benefit from collaboration. What is the point of half a dozen almost identical DEMO plants? With multiple countries now already starting on their own (premature) design for a DEMO or power plant, one actually wonders whether there is any point in continuing with ITER after its construction is completed. The construction will teach engineers many lessons that are useful for building a DEMO, but the ITER design itself dates back thirty years and for DEMO or a real power plant much more advanced technology will be used. The versions that are being designed now cannot benefit much from experiments to be carried out on ITER after 2035 or so, when its most important operations, those with deuterium and tritium, are supposed to start.

However, ITER will probably continue so long as the EU continues to pay almost 50% of the cost, as it guarantees a cheap ride for the other ITER Members. That may change when construction finishes and the EU contribution drops to 34%, and if the cost of running ITER turns out to be higher than currently expected. In view of the different routes the various countries seem to be taking, it will be hard in future to forge an ITER-like collaboration for DEMO or for a real power plant, which, if the EU Roadmap is any guide in this, must be built by or in close cooperation with industry, i.e., private companies, while in view of the complexity and cost of such a project, collaboration between nations would be the only possible way to achieve success. Industry will of course always be prepared to 'contribute' if money can be earned. For operating a nuclear fusion device, it will be hard though to find enthusiasts among currently operating utility companies. They are used to the simple process of burning coal, oil or gas and not likely to be happy running such a complicated device.

European Roadmap to Fusion

Since the alleged successes of JET as the largest tokamak in the world and the efforts to design and construct its successor ITER, the EU has taken the lead in the global fusion effort, in which role it takes excessive pride, so it seems. The European roadmap to fusion covers the entire path from the construction of ITER via a demonstration (DEMO) plant to the final step of a commercial fusion power plant. Most national fusion programmes are based on such a three-step strategy to commercialization. The 2018 *European Research Roadmap to the Realisation of Fusion Energy* sets out in detail how the fusion enterprise will have to be brought to fruition.

The programme has been divided into short-term, medium-term and long-term prospects.

The short term encompasses the construction of ITER; R&D in support of ITER (which is ongoing all over the world); deuterium-tritium experiments in JET, designed to achieve and investigate ITER-like plasma regimes of operation (to start in 2021); conceptual design phase of DEMO (early designs should be ready by 2027); R&D for DEMO; and construction of a dedicated fusion materials testing facility.

The medium term must then see the first scientific and technological exploitation of ITER (so after December 2025 when first plasma is expected); the first exploitation of the fusion materials testing facility; the engineering design phase of DEMO with industrial involvement; the development of power plant materials and technologies; and possible further development of the stellarator concept (see Chap. 14).

And the long term is reserved for high performance and advanced technology results from ITER; qualify long-life materials for DEMO and power plants; finalisation of the design of DEMO; construction of DEMO (to start a few years after high performance deuterium-tritium operation in ITER is achieved, now scheduled to start in 2035); demonstration of electricity generation; commercialisation of technologies and materials; and deployment of fusion together with industry (all this will probably take us into the next century).

Consulting the 2012 document *EFDA Fusion Electricity: A roadmap to the realisation of fusion energy*, it transpires that at any rate in 2012 short term meant up to and including 2020 (now probably a decade later), medium term the decade after (quite a lot of work for just a decade, keeping in mind the slow and delay-prone practice in fusion research) and long term from roughly 2030–2050, so that electricity production from fusion should be available from 2050.

The 2012 EFDA Roadmap states explicitly that "fusion can start market penetration around 2050 with up to 30% of electricity production by 2100"! A stupendous prediction, and most certainly far off! Since ITER is delayed by at least ten years, a few decades should at least be added to these forecasts, but even so the prediction is too audacious to contemplate. To predict that an unproven, far from realised technology will be able to capture 30% of a mature market within such a short time is just beyond comprehension. Global electricity production in 2040 will be around 40,000 TWh (terawatt hours), 40,000 million MWh, assuming that demand for electricity will grow by about 2% per year.[1] It implies that the global electricity generating capacity is about 7 TW. By 2040, if all goes well, ITER will produce just 500 MW of *thermal* power for 400 s. If converted into electricity it is just one third of this, meaning that ITER will still consume more electricity than it produces. It will be a tremendous feat if the DEMO plant, planned after ITER, is capable of breaking even in that respect, i.e., producing as much electricity as it consumes. In any case, it will not be producing more than a few hundred megawatt of electricity. In the 2012 Roadmap it was foreseen to start operation in the early 2040s. This will be at least a decade or two later. Then a pilot plant will have to be built, before construction of any commercial power plants can begin. There is therefore no possibility whatsoever for fusion to start 'market penetration' around 2050. And to predict that around the year 2100 fusion will satisfy one third of global electricity demand, i.e., something like 10,000 million MWh of electric power or in other words have a steady-state capacity of more than 1 TW or 1 million MW, is completely ludicrous.[2] Such predictions have no reasonable basis in fact!

The follow-up to the 2012 *EFDA Roadmap*, the 2018 *European Research Roadmap to the Realisation of Fusion Energy*, is a little more cautious in this respect: "The quest for fusion power is driven by the need for large-scale sustainable and predictable low-carbon electricity generation, in a likely future environment where the global electricity demand has greatly increased. This demand is expected to perhaps reach 10 TW [of global generating capacity] in the second part of this century, by which time the vast majority of energy sources needs to be low-carbon. To make a relevant contribution worldwide, it is estimated that fusion must generate on average 1 TW of electricity in the long-term, i.e., at least several hundred fusion plants in the

---

[1] Global electricity generation in 2018 amounted to 26,700 TWh, of which about 2/3 came from fossil fuels (coal (38%), gas (23%) and oil (3%)).

[2] The capacity factor of a power plant ranges from 30% for wind/hydro plants to 90% for a nuclear fission power plant. A 1000 MW power plant with a 100% capacity factor would generate about 1000 × 24 × 365 MWh = 8.8 TWh of electric power. If the capacity factor is less than 100%, the generated power is correspondingly lower.

course of the twenty-second century." In the course of the twenty-second century is rather vague and can mean a century later than "by 2100" as stated in the 2012 Roadmap. It is hard to say anything sensible over such a long timespan. Such predictions have little value. By that time the climate-change battle will be over, either won or lost. For fusion to be able to make a contribution to this battle it must indeed generate a sizable amount of electricity by 2100. The 2018 Roadmap seems to realise that this is not feasible. As argued above, in the most optimistic scenario the first fusion power plant may perhaps be ready by 2100. An *electricity* (not thermal) generating capacity of 1 TW means 200 5000 MW fusion plants running continuously throughout the year, so it would indeed imply several hundred fairly large plants, for a power station with a nameplate (electricity) capacity of 5000 MW is not a particularly small one. Most nuclear fission power stations currently in operation have much smaller capacity (1300–2000 MW). Moreover power stations don't run continuously during the year. It is of course not known what the corresponding efficiency (called the capacity factor) of fusion power plants will be, but assuming that it is 50% we will need twice the above-mentioned number of fusion power plants, i.e., 400 5000 MW plants, for an electricity generating capacity of 1 TW. So, the long-term European Roadmap seems to have lost all sense of reality. This without mentioning cost and other economic aspects of nuclear fusion power stations, which will be discussed in Chap. 17.

So let us leave aside the long-term predictions and concentrate on the short-term and medium-term technical goals of ITER and a possible DEMO design. Figure 12.1 gives the schedule for this, as included in the 2018 Roadmap.

ITER is the key facility of the roadmap and indeed of nuclear fusion, for without ITER there will be no fusion, although some of the ITER Members continue to take part in the project and at the same time behave as if they can do without. If ITER fails, there will probably never be a nuclear fusion power plant on this earth, at any rate not in another 100 years or so. So there is an awful lot at stake here! The schedule above shows how the progress made with ITER will feed into DEMO, with the corresponding periods when this is supposed to happen. In this schedule DEMO is to last until 2060, which confirms the above assessment that fusion will not make a significant contribution to carbon-free energy generation in this century.

The problems identified in the previous chapter are also recognised in the Roadmap, which views a power exhaust system (divertor) capable of withstanding the large thermal loads, the creation of a dedicated neutron

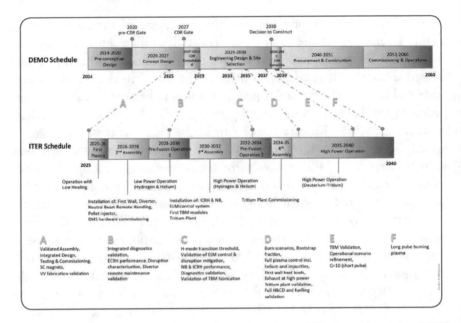

Fig. 12.1 Diagram depicting how information from ITER flows into the DEMO Conceptual and Engineering Design Activities. CDR = Concept Design Review

source for materials research, and the self-sufficiency of tritium as the main challenges.

A surprising point in the Roadmap is that the stellarator is seen as a possible long-term (whatever that may mean) alternative to a fusion power plant based on the tokamak. It shows that confidence in the tokamak concept is no longer universal, unless there are political reasons for including the stellarator here. Germany currently operates the largest stellarator: Wendelstein 7-X, which will be discussed in Chap. 14. No other country involved in ITER or otherwise has made room in its research programme for stellarators as a viable alternative to the tokamak. One of the main reasons for pursuing the stellarator line arises from the fact that they operate without a large plasma current, resulting in an inherent steady-state capability and the absence of plasma disruptions, which are both challenges for the tokamak. Stellarator development is, however, far behind the tokamak and, if it were adopted as the concept for a power plant, several decades would have to be added to the roadmap schedule based on the tokamak. A stellarator-based DEMO or power plant is out of the question in this century.

The Roadmap further states that industry must be involved early in the DEMO definition and design. The evolution of the programme requires industry to progressively shift its role from being a provider of high-tech components to a driver of fusion development. It is not made clear which industry is meant here. Surely not the industry that builds the components for ITER, DEMO or power plant, as they will not be interested in running a power station. Utility companies normally run power stations, but they do not know anything about nuclear fusion, neither are they at present involved in the construction of ITER. Fusion has so far been a fully government-sponsored research effort. The only industry one could remotely think of is the nuclear power industry, and hence the companies building nuclear fission power stations, but they too know nothing about fusion. Nuclear fission is quite a different kettle of fish. Industry, the Roadmap continues, must be able to take full responsibility for commercial fusion power plants after successful DEMO operation. For this reason, DEMO cannot be defined and designed by research laboratories alone, but requires the full involvement of industry in all technological and systems aspects of the design. I don't see ordinary utility companies contributing anything to the design of DEMO, nor do I see them running a nuclear fusion power station or taking full responsibility for it. As soon as DEMO or another power plant is connected to the grid, it will sell the generated electricity to a utility company, I suppose, which will then sell it on to the consumer, but it will not itself run the nuclear fusion plant. A new industry will have to be built up for this with very different expertise than utility companies currently possess and, in view of the exorbitant cost and risk of a nuclear fusion power plant, with heavy involvement and far-reaching guarantees from governments.

Plans of the Various ITER Members

Late in 2013 the ITER Members were already presenting their own projects for DEMO. All have the objective of building their machine by 2050. The various Members are exploring different routes towards DEMO and some different regroupings have taken place, whereby some countries (Russia, South Korea, India) seem to want to go it mostly alone or not at all (US). Japan, Korea, India, Europe and Russia stated their intention to begin building DEMO in the early 2030s in order to operate it in the 2040s. Such schedules seem a little too ambitious at the present time.

The conceptual designs all sketch out a machine that is larger than ITER with a major radius ranging from 6 to 10 m. In comparison, ITER's major radius is 6.2 m and JET's half of that. The ambitions for DEMO differ. As far as power is concerned the designs vary from an electricity output of around 500 MW for the European DEMO to 1500 MW for the early Japanese

DEMO (which has now been scaled down). For some Members, DEMO will be a pre-industrial demonstration reactor; for others, it will be a quasi-prototype that requires no further experimental step before the construction of an industrial-scale fusion reactor. The Russian DEMO project stands out from the others, as it is a hybrid combining fission and fusion within the same device.

Europe and Japan are cooperating on DEMO design work as part of the Broader Approach agreement. Within that framework they are also building the JT-60SA tokamak and IFMIF. Construction of DEMO is not part of the Broader Approach agreement, which does not go beyond the ITER construction phase. ITER serves as the school where physicists and engineers will learn how to build DEMO. Joint construction of DEMO by Japan and the EU is not foreseen at present.

In 2018, Japan reviewed its strategy for developing a DEMO fusion reactor. The new guideline states that the decision to proceed with a DEMO reactor phase will be taken in the 2030s when ITER will demonstrate D-T burning plasmas, with the added condition that the economic feasibility of a commercial reactor must then be foreseeable. One of the objectives of its DEMO reactor will be self-sufficiency in tritium, and this while Japan has so far completely eschewed working with tritium. The Japanese timeline is depicted in Fig. 12.2.

In this timeline, the success of ITER has been assumed, as well as steady-state operation of JT-60SA. Without these assumptions being fulfilled, there will be no point in starting construction of the DEMO plant. The timeline is clearly overambitious, as it is completely unrealistic to have both a DEMO reactor and a commercial power generating device ready in just ten years, namely by the middle of this century, ten years after ITER has achieved its objective.

Although the decision on DEMO has now been postponed to the 2030s, Japan is nevertheless currently working on the design of a steady-state JA DEMO with a major radius of 8.5 m, minor radius 2.42 m and *thermal* fusion power output of 1.5–2 GW (three or four times the value for ITER), including also a tritium-breeding blanket and niobium-tin superconducting magnets like ITER. The conceptual design stage lasts until 2027, then follows an engineering design stage until 2032, and after that the manufacturing design stage until 2036. This DEMO, which does not even use high-temperature superconducting magnets, will be out of date before construction starts.

According to the European Roadmap DEMO will mark the very first step of fusion power into the European energy market by supplying between

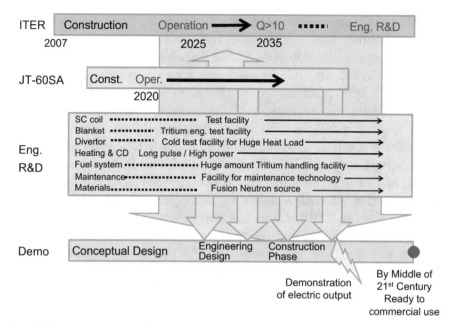

Fig. 12.2 Schematic timeline of Japan's roadmap towards DEMO and commercial plants

300 and 500 MW of net electricity to the grid, which is approximately 3–4 times less than an average nuclear fission reactor. To achieve this electricity output, net fusion power must be 1–1.5 GW. Europe's DEMO would be a "demonstration power plant" to be followed by the first-of-a-kind fusion power plant. It will largely build on the ITER experience and will have to breed its own tritium; demonstrate materials capable of handling the flow of neutrons produced in the fusion reactions; and demonstrate safety and environmental sustainability, and sufficient technology to allow a first commercial power plant to be constructed. Very little of this has been accomplished and the experience with ITER, although useful, will not help that much. The demands differ very little from the Japanese demands for DEMO.

Following the delays in ITER, early conceptual design(s) of a European DEMO are expected around 2027 and construction is now planned for after 2040. After high power burning plasmas have been demonstrated in ITER, DEMO will be operational for around 20 years, i.e., until after 2065, if there are not too many more delays. To achieve fusion electricity in the second half of the century, European DEMO construction has to start in the early 2040s, shortly after ITER is supposed to have achieved its $Q = 10$ milestone, if it ever does so. Therefore, engineering design for a DEMO will become a major activity after 2030.

Within the framework of SIFFER (Sino-French Fusion Energy Center) China and France have joined forces and made a schedule for the China Fusion Engineering Test Reactor (CFETR), which aims to bridge the gap between ITER and DEMO. Its main objectives are to complement ITER; produce 200 MW of fusion power (less than ITER); demonstrate a full tritium fuel cycle, with a tritium-breeding ratio above 1.0; explore options for remote handling techniques; and address the physical and technical solutions for achieving steady state operation. It took nearly four years to complete the conceptual design of CFETR, which was achieved in 2017, showing that China is in a hurry. A preliminary estimate of the total project cost of the fully superconducting tokamak is about $4 billion, an astonishingly low figure when compared to ITER. China has now progressed to the detailed engineering stage of CFETR and construction is planned to start in the early 2020s. It will already be upgraded to a full DEMO in the 2030s to obtain a fusion power production of 1000 MW, a fusion gain (Q) higher than 12, and tritium self-sufficiency. This programme implies that China's DEMO might be (close to) working before ITER has even started its D-T operations. In any case it shows that at least for China ITER is apparently not essential for designing a DEMO reactor or for running its own D-T operations, although it has so far no experience at all with tritium. The construction of a Prototype Fusion Power Plant (about 1 GW of electric energy) will then be completed in 2050–2060 and be the final step in the Chinese roadmap towards a commercial fusion power plant. In summary, China, like Japan, is already planning its own sequel to the ITER adventure, both countries apparently expecting that, after ITER, they have to go further on their own (in China's case with some help from France), but apparently the Chinese want to start building their demo plant even before ITER is ready!

According to a recent report, India's energy demands are rising rapidly and to meet future requirements, a development strategy for a demonstration power reactor (DEMO) is planned. According to the Indian Roadmap of Fusion Energy, an intermediate fusion machine is planned before DEMO development, viz., the Steady State Tokamak-2 (SST-2). It will be a D-T machine with the aim to test and develop components for a demonstration plant around 2027. It is a medium-sized device with low fusion gain ($Q \sim 3$–5). The fusion power output can be from 100 to 300 MW. It is apparently thought that ITER ($Q = 10$ and a similar power output) cannot answer all the questions and SST-2 is required to fill the technological gaps between ITER and a demonstration plant. One of its objectives is to show the capability of a tritium fuel cycle, which is beyond the scope of ITER. Construction of a demonstration plant will then start as soon as 2037. The

SST-2 is still under development and the situation with DEMO is difficult to assess, but it is clear that India wants to pursue an independent route towards such a device and has currently no intention of starting any collaboration beyond ITER.

South Korea initiated a conceptual design study for a K-DEMO in 2012, with as target the construction of the device by 2037 with potential for electricity generation starting in 2050. In its first phase (2037–2050), K-DEMO will demonstrate a self-sustained tritium cycle, and develop and test components. Then, in the second phase after 2050, a major upgrade is planned, in order to show net electricity generation of the order of 500 MW (roughly 1500 MW of thermal fusion energy). The main parameters of K-DEMO have already been defined. It is expected to be a 6.65 m major radius tokamak, a bit larger than ITER.

Russia plans the development of a fusion-fission hybrid facility called DEMO fusion neutron source (FNS), a conventional tokamak (fairly small with major radius $R = 3.2$ m) with a superconducting magnet system that would harvest the fusion-produced neutrons to produce fissile material (through interaction of the neutrons with heavy elements like thorium or depleted uranium) and to break down radioactive waste. Especially the latter application is interesting as it could be used to promote a revival of ordinary nuclear fission plants by (partly) solving the nuclear waste problem. It can also be used for the production of tritium. Development of the device started in 2013 and construction is expected in 2033. It is part of what is called Russia's fast-track strategy to a fusion power plant by 2050.

If the DEMO-FNS project is successful, an experimental industrial scale hybrid facility (Pilot Hybrid Plant) with a fusion power of 40 MW and total thermal power of 500 MW (including the fission part) will be built to demonstrate the possibility of producing electric power from nuclear fuel and offering radioactive waste disposal services on a commercial scale.

The US has not yet officially engaged in a DEMO project, but plans have been developed for a so-called Fusion Nuclear Science Facility as an intermediate step between ITER and DEMO to be used for developing and testing fusion materials and components for a DEMO-type reactor. Plans call for operation to start after 2030, and construction of a DEMO after 2050. It is unlikely though that the US government will provide funding for these projects.

The main impression you get from all this is that ITER actually does not matter that much. Most DEMO plans and designs seem to be able to do fine without ITER. None of the problems identified in the preceding chapter can stymie the enthusiasm for forging ahead. Experience with working with tritium does not seem to be necessary, tritium self-sufficiency seems to be a piece of cake, and any problems will be resolved as we go along.

Power Plant Studies

Designing real nuclear fusion power stations has already been a pastime for decades. The early argument given in the 1960s was that the question of power plant design is as important as the question of plasma confinement, since what is the good of plasma confinement if we don't have any idea whether a reactor system is feasible? Although that may be a valid question, it was equally clear that without plasma confinement there would be no nuclear fusion and, moreover, the question of the feasibility of a nuclear fusion power plant could not be settled in those early days.

September 1969 saw the first international conference on fusion reactors held at Culham in the UK. This date is traditionally taken by the fusion community as marking the beginning of sustained and serious interest in fusion reactor design studies and of the encroachment of nuclear engineers upon the fusion territory. In such designs no hardware is produced, but design work is carried out with computer codes. Since this conference, more than 50 conceptual power plant studies have been conducted in the US, EU, Japan, Russia and China.

Desirable characteristics of a practical fusion reactor were already listed in a NASA technical memo in 1976 and include:

a. Steady-state operation (power interruptions associated with cyclic operation are awkward for utility companies).
b. High *beta* (implies smaller reactor size, lower capital investment, e.g., in the magnet system).
c. Self-sustaining fusion reaction (without this characteristic commercial power generation would be impossible).
d. Possibility of advanced fuel cycles (meaning other fuel than D-T, since such fuel cycles release more of their energy in the form of charged particles).
e. Direct conversion into electric power (related to (d), as the energy of charged particles can be converted into electric power by direct conversion schemes).
f. No neutrons or activation of structure (related to (d) and (e); the advanced fuel cycles either do not generate neutrons or at least minimize the neutron generation and/or activation of the reactor structure).

g. Environmentally safe (i.e., minimizing possible radiation hazards).
h. High capital and resource productivity (capital expenditure should not be too high and the return on investment should be adequate).

Desirable does not of course mean that all these characteristics are necessary and it will not be possible to design a plant that has all these characteristics, but it is striking that the tokamaks (spherical or otherwise) that have been or are currently being constructed, do not naturally meet a single one of these criteria, and nonetheless it is virtually the only approach that is being pursued.

In the 1970s a group at the University of Wisconsin was especially active in the field of reactor design and turned out various tokamak reactor studies. In total this group was actively involved to a greater or lesser extent in 44 magnetic fusion energy designs.

From around 1990 the ARIES program (Advanced Reactor Innovation and Evaluation Study) in the US was leading in designing fusion reactors. It was a multi-institutional national research project at the University of California at San Diego, with among others involvement from Wisconsin, to conduct advanced fusion systems research and to explore the potential for fusion development. It continued working until around 2013. The purpose of the programme was to develop fusion reactors with enhanced economic, safety, and environmental features. The programme produced a whole series of conceptual studies of fusion systems. When one design was finished, they could start again to keep up with the progress that had in the meantime been made in physics and technology. It is interesting to note that, as new physics and new technology became available, the ARIES designs became progressively smaller and consequently cheaper.

Practical considerations such as high safety, environmental friendliness, public acceptance, reliability as a power source and economic competitiveness were paramount considerations in these studies, which are in general very detailed, putting a value on every component of the plant and even (quite preposterously) predicting the price per kWh of the electricity generated. They give the feeling that it is just a matter of choosing a design from the shelf and building a reactor.

A pertinent remark, applying to all designs, is that the physics of burning plasmas and α-particle dynamics remain unresolved issues in fusion research. Obtaining advanced tokamak modes in the presence of dominant α-particle heating is a critical issue that can only be addressed in a long-pulse, burning plasma experiment, which so far has not seen the light. In other words, not enough is currently known to design a nuclear fusion power plant with any confidence, but it is nevertheless done with all the confidence in the world.

Apart from two early Russian designs, no design studies were carried out outside the US until the 1990s and, not counting the DEMO designs, there are in total only a dozen with the majority coming from the fossil-fuel starved countries in the Far East, reflecting a different attitude towards nuclear fusion. While US studies ended about fifteen years ago, the rest of the world is still busy designing DEMOs and other reactor types.

As regards waste management, a European study carried out in the 1990s concluded that over their lifetimes, fusion reactors would generate, by component replacement and decommissioning, activated, i.e., radioactive material similar in volume to fission reactors, but qualitatively different in that the long-term radiotoxicity is very much lower. After about a hundred years (still a considerable time), radiotoxicity indices (relating to ingestion and inhalation) for the total activated materials fall to levels comparable with the ashes from coal-fired plants. The study indicates that fusion waste would not constitute a burden for future generations. The inclusion of the latter indicates, in my view, that some politics was involved, as such a conclusion can in no way be scientifically justified. We will come back to this later in this book, where it will also be pointed out that the volumes of radioactive material from a fusion plant are vastly bigger than those from a fission plant.

The EU Power Plant Conceptual Study from the beginning of this century focussed on five models that spanned a wide range of physics and technology options. All these models produce about 1.5 GW of electricity, but like the ARIES models, become progressively smaller and use less power. Safety and environmental issues were carefully considered. The study highlighted the need to establish the basic features of DEMO, to bridge the gap between ITER and the first-of-a-kind fusion power plant. It also gave an estimate, for what it is worth, of the cost of electricity generated by fusion, and concluded (in 2004) that fusion compares favourably with other renewable sources, such as solar and wind. The price of electricity from solar and wind has come down a lot in the last fifteen years, making it the cheapest available, while fusion has only piled on the costs.

Japan has been quite active in power plant design work since the early 1990s. Including the recent DEMO designs, it conducted 11 design studies, which were not limited to tokamaks.

A series of pioneering steady-state tokamak reactor (SSTR) studies was developed in Japan in the 1990s to achieve high power density through high magnetic field strength.

China too has conducted a series of design studies in the last 15 years, covering a broad range of tokamak concepts.

Conclusion

It may be obvious, but I feel that it is still worth noting that power plant design studies have now been carried out for more than half a century, with most design reports ending by saying something like: "results for the near-term models suggest that a first commercial fusion power plant - one that would be accessible by a "fast track" route of fusion development, will be economically acceptable, with major safety and environmental advantages." This citation originates from a 2007 paper on the EU Power Plant Conceptual Study, but equally optimistic and unwarranted conclusions can be found in a multitude of other reports. It goes almost without saying that a considerable gap remains between the performance required in these designs and that obtained in the laboratory to date. It will still take probably at least half a century before the first fusion power plant will see the light (if it ever happens). By then a century will have passed since the first design. This fact alone is probably sufficient for the reader to draw their own conclusion, and it certainly does not bode well for the future of energy from nuclear fusion.

13

Spherical Tokamaks

Spherical tokamaks (STs) are characterised by a small(er) central hole in the doughnut than conventional tokamaks, resulting in a smaller aspect ratio $A = R/a$, with R and a the major and minor radius, respectively. In this chapter we will briefly discuss what the advantages and disadvantages of such a geometry are or could be. Interest in this design is on the upsurge and most new tokamaks built in this century are actually spherical tokamaks. The reasons for this are that they can still be built rather cheaply and, more importantly, bring some new insights.

In a spherical tokamak the size of the doughnut hole is reduced as much as possible, resulting in a plasma shape that is almost spherical with little room left for a central column nor, more importantly, for the inner limbs of the toroidal field coils (the blue coils in Fig. 7.4). For STs, aspect ratios are equal to or smaller than 2, with extreme ones as small as 1.1. Figure 13.1 shows the difference between the plasma shapes for an ST and a conventional tokamak with high aspect ratio.

A smaller central hole is essentially the only difference. This seemingly simple change has profound consequences, which, so its advocates would have us believe, could open up a faster route to a fusion power plant, without the massive scale and cost, than the route followed via ITER. As a reminder, a large solenoid normally runs through the hole down the centre of a tokamak. See Figs. 7.2 or 7.4 to recall the basic setup of a tokamak. By varying the electrical current in the solenoid, a current is induced in the plasma. This current heats the plasma and contributes to the poloidal magnetic field which,

L. J. Reinders, *Sun in a Bottle?... Pie in the Sky!*,
https://doi.org/10.1007/978-3-030-74734-3_13

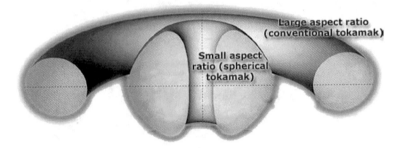

Fig. 13.1 Plasma shape of a spherical (low aspect ratio) tokamak compared to a conventional (high aspect ratio) tokamak

together with the toroidal (and other poloidal) fields created by external coils, completes the forces that confine the plasma.

So why would a compact or spherical tokamak be advantageous or preferable? Potential advantages of this shape were first suggested in the second half of the 1970s. Based on a large number of calculations of plasma instabilities in tokamaks with aspect ratios ranging from 2.5–5, it was shown that the maximum stable *beta* value (the ratio of the thermal plasma pressure to the magnetic pressure) increases with decreasing aspect ratio and with increasing elongation. Just decreasing the aspect ratio by a factor of 2 would double the value of *beta*! A high *beta* is a desirable feature and, with other factors being constant, such an increase would be a considerable and important improvement for magnetic confinement fusion. Conventional tokamaks operate at relatively low *beta*, 4 or 5%, with the record being just over 12%. Calculations have shown that practical designs would need *beta* values as high as 20%, which would be easily achievable in an ST. The higher *beta*, the higher the fusion power, since fusion power rises as *beta* squared. At the same time STs need less toroidal field, the plasma volume can be reduced, and the fusion power can be produced in a smaller device. So, an ST would mean high power density in a small device.

When these considerations are worked out further, it turns out that the vertical (poloidal) field needed in tokamaks has the natural tendency to elongate the plasma and that this elongation increases as the aspect ratio goes down, so reinforcing the increase in *beta*. As Fig. 13.1 shows, the cross-section of the plasma in an ST is elongated in the vertical direction, while its magnetic topology remains virtually the same as that of a conventional tokamak, with a toroidal field generated mainly by toroidal field coils and a poloidal field generated by the plasma current and poloidal field coils. If the aspect ratio goes down from 2.5 to 1.2, the elongation, indicated by κ, increases from 1.1 to 2, and the toroidal magnetic field required to give the same quality or safety factor q for a given plasma current falls by a factor of 20! This implies that the value of *beta* also increases by a factor of 20.

In summary, the low aspect ratio and large elongation imply that the *beta* values that can be reached by STs are very high (about 40%). Improving *beta* means that less energy is needed to generate the magnetic fields for any given plasma pressure (or density), and reactors operating at higher *beta* are less expensive for any given level of confinement.

With the central hole of the torus squeezed as much as possible, there is, however, little room left to fit in all the necessary equipment for a central solenoid, needed if a current is to be inductively driven, for the shielding it requires, and for the inner limbs of the toroidal magnetic field coils, which, if superconducting material is used for them, likewise require extensive shielding. This implies in the first place that in an ST configuration other current drive methods must be applied, something other than through induction from a central solenoid. Important for the viability of the spherical torus concept is that progress has been made in alternative (non-inductive and other) current drive schemes, also inspired by the fact that an eventual reactor must be able to operate in a steady state, and not as a pulsed device. Several forms of non-inductive start-up are presently under active investigation.

The spherical tokamak idea is still very young, although a few STs date from quite a while back, starting with the Japanese Asperator T-3 from 1974. It was followed by the Erasmus tokamak in Belgium, dating from 1976 and vying for second place with the South African Tokoloshe tokamak. A large number of spherical tokamak devices have since been built and are or have been operational in more than a dozen countries. The website www.tokamak. info gives a list of about 40 STs. So far no fusion power whatsoever has been produced in any such device. Their development is still some years, perhaps decades, behind conventional tokamaks, but since spherical tokamaks are much smaller and cheaper, they can be built quickly and could catch up fast.

The construction of such devices really took off from the early 1990s after the START (Small Tight Aspect Ratio Tokamak) device at the Culham Centre for Fusion Energy (CCFE) at the JET site in Culham in the UK had shown that the theoretical advantages of STs as explained above hold up in practice. The Culham centre has pioneered the spherical tokamak fusion concept and still plays an important role in its current development. JET had already reduced the aspect ratio to around 2.4, but it was the START device that revolutionized the tokamak by changing the previous toroidal shape into the tighter, almost spherical, doughnut shape and in the subsequent experiments verified the aforementioned advantages of STs (high elongation, high *beta*). START and its successor MAST (Mega Ampere Spherical Tokamak), plus

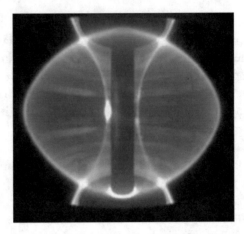

Fig. 13.2 Camera image of a plasma in the MAST reactor. Note the almost spherical shape of the outside edge of the plasma. The high elongation is also evident. The central post is part of the toroidal field coil

their American equivalent NSTX, have provided the most information on STs and by doing so they were the instigators of the current, still modest, upsurge in spherical tokamak research.

Figure 13.2 shows a beautiful picture of an almost spherical plasma in MAST.

An upgrade of MAST (MAST-U) is currently being completed. A little premature in my view, since MAST-U has not yet been shown to work, but in line with the trend in fusion research, a follow-up of MAST-U is already being planned in the form of STEP (Spherical Tokamak for Energy Production), which as its name indicates is rather ambitious, going even beyond the future DEMO for conventional tokamaks. The STEP programme aims to deliver an integrated concept design for a fusion power plant based on the spherical tokamak, and develop and identify solutions to the challenges of delivering fusion energy. Its equally ambitious technical objectives include the construction of a fusion power plant by 2040 (invoking so it seems the long-established magical two-decade rule for fusion power to be realised) based on a design that should be ready in 2024.

The UK government has announced £220 million of funding over four years for the conceptual design of STEP. With this the UK essentially turns its back on the conventional tokamak, having (rather suddenly) placed all its eggs in the spherical tokamak basket. The programme is still at a very early stage with £20 million having been made available for the first year in September 2019, but they intend to move fast and in December 2020 communities in the UK were already invited to volunteer a site for the reactor.

Moreover, in November 2019, with the ink on the first announcement barely dry, the government announced further extra funding of £184 million for the next five years for MAST on the condition that it will contribute to STEP. This amount comes on top of the £220 million mentioned above. A further £86 million was promised for a Nuclear Fusion Test Facility (NFTF). So, in total the UK government has now committed £490 million for STEP studies, material studies, etc., bravely starting on the march towards power generation by nuclear fusion using spherical tokamaks.

In the United States, after the shutdown of TFTR, PPPL also shifted its attention to spherical tokamaks and built NSTX (National Spherical Torus Experiment). NSTX and its upgrade belong to the most powerful STs in the world. First plasma was obtained on NSTX in February 1999. Its parameters were very similar to those of MAST, with aspect ratio 1.5, $R = 0.85$ m and $a = 0.68$ m, so a very small device compared to JET or ITER.

About 60 percent of the total plasma current in NSTX was obtained from alternative current drivers, relaxing the need for induction (from a central solenoid) to sustain the current. Although important, this had already been achieved at TFTR and does not make the ST configuration stand out. NSTX also copied START's record of a toroidal *beta* of about 40%. The energy confinement time in NSTX has been consistently 1.5–2.5 times longer than expected from the results accumulated from conventional tokamaks that were run in their basic mode of operation. This is a very favourable result for the ST's potential in fusion energy development.

In 2012 NSTX was shut down as part of an upgrade program and became NSTX-U. The upgrade boosted the capabilities of the NSTX reactor and made it the most powerful spherical tokamak in the world. It has doubled the field strength to one tesla, and the electric current flowing in the plasma has been increased to 2 MA.

NSTX-U was completed in the summer of 2015. First plasma was subsequently achieved and H-mode very quickly accessed. Just after its first 10 weeks of running, in the summer of 2016, NSTX-U had a mishap (a 'significant technical issue') when one of the machine's 14 magnets, a poloidal field coil, shorted out. The mishap cost the PPPL director his job and the expected one year delay has now stretched to more than four and the upgrade is not expected to run before 2022. The reason for this long delay is not clear, but it shows the sensitivity of such devices. A seemingly small thing goes wrong and the entire machine is out of commission for a very long time.

A Next Step Spherical Torus (NSST) experiment has been proposed and, by taking advantage of the tritium experience with the former TFTR facility at PPPL, it is to test the ability of the ST to confine deuterium-tritium fusion-producing plasmas. It is envisaged to play a role for STs similar to that of JET, JT-60, and TFTR for conventional tokamaks. Although studies are underway for the design of an NSST experiment, there are currently no plans to provide public funding for a spherical tokamak or anything else beyond the NSTX-U; a sensible attitude as the NSTX-U is not even running.

Most public money made available nowadays for fusion research is drained away by the insatiable demands of ITER or by devices that are mobilised to test parts of the ITER design. For a decade or so, private companies, mainly start-ups and funded by venture capital, have been involved in (spherical) tokamak research. As far as STs are concerned, the most important company is the British **Tokamak Energy** (founded in 2009), which is undoubtedly the leader in the private ST field. It has so far been a fairly small player on the fusion playing field of private capital, but is slowly building up its position through a well-thought-out research and construction plan. The company grew out of the Culham laboratory. Some of its leading scientists were heavily involved first with START and MAST and later with JET, experience they now use to build their own fusion devices.

In 2019 they were still claiming to produce net fusion power just after 2025, while the latest version of the website makes the heavenly promise of clean and abundant fusion power by 2030. The words 'clean' and 'abundant' are the most questionable. Fusion power will certainly not be abundant by 2030 and neither will it be clean. Another page of the same website just speaks rather modestly of demonstrating "the feasibility of fusion as an energy source by 2030". Such partly misleading statements seem to be one of the industry's characteristics. It is hoped that the tough-minded venture capitalists also take it lightly.

The company has set out building a series of spherical tokamaks and started in 2012 with the first version of its **ST25**. ST stands, quite unimaginatively, for Spherical Tokamak. Its major radius was, again not surprisingly, 25 cm. Unusually, the whole table-top device was contained in a small room and powered from only a 32A 415 V supply, i.e., ordinary three-phase electric power.

The second version of the machine, **ST25-HTS**, which is now in the Science Museum in London, used high temperature superconducting magnets and provided the world's first demonstration of a tokamak magnet system where all the magnets are made from high-temperature superconductors (HTS).

The advantage of HTS is that, in addition to operating at relatively high temperatures, they can also produce and withstand relatively high magnetic fields. The coils are cooled by helium gas to 20–50 °K (−253 to −223 °C). Conventional low temperature superconductors (LTS) have to be cooled to a few degrees above absolute zero and would need thick shielding (of about 1 m) to prevent neutrons from heating the superconductors. This shielding would have to be installed in the central column and would make the device extremely large. HTS promises a solution to this problem, although the necessary shielding will still increase the aspect ratio.

HTS is not all roses though. In the first place there is the increased pressure the coils are producing on themselves, because the fields are much higher. HTS magnets, which are quite differently constructed and much more compact, are, however, much stronger than LTS magnets. A second disadvantage is the vastly increased neutron wall load. The ST devices constructed by Tokamak Energy are many times smaller than ITER, but the power density and neutron load will be similar, e.g., the divertor loads are equal.

STs have so far operated at toroidal fields of less than or just equal to 1 T. For high fusion performance, devices operating at 3 T or above are needed. This will require innovative engineering solutions especially for the central column. To develop and demonstrate such solutions, Tokamak Energy has recently constructed the **ST40** (Fig. 13.3 shows the full device) which has copper magnets and is intended to operate at magnetic fields up to 3 T. It will not be equipped with HTS magnets, but its successor devices will. ST40's main objective is to demonstrate the feasibility of the ST concept with such high fields, aiming at plasma temperatures in the 10 keV (100 million degrees) range.

The experiments on this machine will start somewhere between START and MAST, and will end at parameters of up to a factor of ten higher than MAST, i.e., burning plasma conditions, not enough to obtain fusion of the actually used deuterium-deuterium plasma, but sufficient when extrapolated to deuterium–tritium. If this is successful, it will have been shown that fusion conditions can be achieved on a comparatively small machine, much smaller than the big tokamaks like JET and TFTR, let alone ITER.

Compared with the large STs discussed above its physical dimensions are considerably smaller than MAST and NSTX, but its toroidal field is much higher. With this the ST40 is the first high-field spherical tokamak and a necessary step towards a steady-state ST reactor. The plasma inside the tokamak will reach more than 100 million degrees, and this while just 40 cm away the copper coils are cooled with liquid nitrogen to −196 °C, a temperature gradient that is unparalleled. With HTS magnets the gradient will be larger still.

Fig. 13.3 The vacuum chamber surrounded by copper toroidal field coils on Tokamak Energy's ST40 spherical tokamak

The experience obtained with the low-field HTS device ST25-HTS and the high-field ST40 will be used to design and construct a high-field ST with HTS magnets. The main objectives of the successor devices, named ST-F1 and ST-E1, are to demonstrate efficient production of neutrons and scientific breakeven with an upgraded neutral-beam injection system in deuterium-deuterium plasmas and in deuterium-tritium plasmas subject to site availability and a licence to work with (radioactive) tritium. Tokamak Energy intends to move fast and to have at least the ST-F1 running before 2025. It is already clear from this date that the promise of "abundant fusion power by 2030" is empty. Why are such foolish statements made?

The ST-F1 will be a demo device (major radius probably 1.4 m) and its successor the ST-E1 will be a power plant model (major radius probably 2 m). It should be clear that these devices will be vastly smaller than ITER. The proposals are still on the drawing boards and the feasibilities of the devices depend critically on obtaining satisfactory solutions for some critical engineering problems which are similar to the problems encountered with conventional tokamaks, the ones discussed in Chap. 11.

In many respects the ST is now equal to existing medium-sized conventional aspect-ratio tokamaks, which together with the large devices JET,

JT-60U and TFTR formed the basis for the design of ITER. The ST equivalent of these large tokamaks is still lacking, another reason why it may be somewhat premature to talk about a power-plant design based on the ST concept. There have, however, been developments, in particular high-temperature superconductors, which according to some have also greatly changed the playing field in respect of power plant designs based on conventional high aspect ratio tokamaks.

But size is not the only thing that plays a role. Because the energy confinement times of tokamak plasmas scale positively with plasma size, Lawson's fusion triple product is also generally expected to increase with size. This has been one of the reasons why tokamak devices have become increasingly bigger.

As we have seen, tokamak plasmas are also subject to operational limits, e.g., a density limit and a *beta* limit. When these limits are taken into account, the triple product becomes almost independent of size; and depends mainly on the fusion power. In consequence, the fusion power gain, which is closely linked to the triple product, is also independent of size. Further, it has been found that the triple product is inversely dependent on *beta*. This implies that the minimum power to achieve fusion reactor conditions is driven mainly by physics considerations, especially energy confinement, while the minimum device size is driven by technology and engineering considerations, such as wall and divertor loads. These latter aspects are evolving in the direction of making smaller devices feasible.

So, when the density limit is taken into account, the relationship between fusion power and fusion gain is almost independent of size, implying that relatively small, high performance reactors should be possible, perhaps with a major radius of 1.5–2.0 m (where ITER, not even a reactor, has a major radius of 6 m), a volume of 50–100 m^3 (ITER's plasma volume being bigger than 800 m^3) and operating at relatively low power levels, 100–200 MW. The lower power requirement is especially advantageous.

There are, however, a number of critical engineering problems for which solutions must be found before an ST power plant will be feasible. Many of these challenges apply equally to conventional large devices. They include handling the stress in the central column while simultaneously accommodating the HTS toroidal field magnet; handling the plasma exhaust in the divertor region where power loads will be at the limit of available materials; providing the inboard shielding needed to protect the HTS tape from the intense neutron and gamma radiation for it to have an acceptable lifetime; and reducing the neutron heating to a level that can be handled with a reasonable cooling system.

The advocates of spherical tokamaks claim that STs could result in more economical and efficient fusion power and that designs demonstrating the feasibility of ST power plants have already been developed. The latter is a bit far-fetched as so far no fusion power at all has been produced by an ST device; no breakeven of any form, not even 'scientific breakeven' has been shown. Nor has any route towards a power plant based on the ST configuration been laid out with any clarity. In this respect it is surprising that the UK on its own has launched a £200 million design effort for a fusion power plant based on the spherical tokamak (the STEP programme discussed above), while at the same time still being involved in ITER or ITER-related programs. But the doubt in ITER is rising. Does anyone really have confidence that it will work and that it is a viable route towards *affordable* fusion power? When doubt creeps in, the alternative suddenly looks much better, even if it has nothing to show for itself yet. It offers an open field for a good salesman.

Although STs have unmistakable advantages (compact device with plasmas confined at higher pressures for a given magnetic field; the magnetic field up to ten times less than in a conventional tokamak; much lower cost for the same performance), the disadvantage of the spherical tokamak is also clear. The need for a slender centre column imposes severe engineering issues and space constraints. There is hardly any space for the inner limbs of the toroidal magnets, and because of problems with the shielding of sensitive components against damaging neutrons produced by fusion events, many STs are forced to decrease the minor radius and hence make the aspect ratio of the device bigger. Using low-temperature superconductors (as in ITER) would require a neutron shield thicker than a metre and make a spherical tokamak unfeasible, although this could be remedied by making the device bigger.

A note of caution is appropriate, however. When reading through papers reporting the results of research on spherical tokamaks, you do not get the feeling that any real progress is being made. They grapple with the same problems as conventional tokamaks did before or still do, e.g., instabilities, only at a still lower parameter regime. Temperatures are still low and there is hardly any mention of the Lawson condition, for instance, which has to be fulfilled for STs, too, before any fusion can become a reality. All the research reported so far has not really brought 'power on the grid' from a spherical tokamak much closer.

14

Stellarators and Other Alternative Approaches

A bewildering multitude of fusion approaches has been tried in the past. Most were abandoned early on or swept away in the tokamak stampede, unleashed after the positive results with the Russian T-3 tokamak. The stellarator, which as mentioned in Chap. 12 is considered in the European Roadmap as a possible long-term alternative to the tokamak, was just as promising as the tokamak, but was abandoned all the same, although some pockets of resistance held out in anticipation of better times, as will be discussed below.

Not only is the stellarator making a comeback in the twenty-first century, but we are witnessing a revival of several old practices, tried and abandoned in the past, and most, if not all, will suffer the same fate again. All these approaches focus in one way or another on nuclear fusion, but very few actually make the pretence of being useful as power generating devices. The majority are used for other purposes or just as research tools to study problems in plasma physics. However, tall stories repeatedly appear in the media with claims that one or other such approach is on the brink of a breakthrough and will provide the usual 'clean and unlimited energy' within a decade or so. So far nothing spectacular has come to the fore.

Stellarators

In Chap. 6 we discussed the early stellarator designs studied at PPPL by Spitzer and his co-workers. Stellarator development at PPPL ended with the tokamak stampede. Although abandoned at PPPL, the stellarator idea has not

© The Author(s), under exclusive license to Springer Nature
Switzerland AG 2021
L. J. Reinders, *Sun in a Bottle?... Pie in the Sky!*,
https://doi.org/10.1007/978-3-030-74734-3_14

lain completely dormant. Several groups in various parts of the world maintained faith in this type of machine and have continued their experiments. They are making progress but are still lagging behind the tokamak, at any rate in size, but also as regards the stage of development. Nobody engaged in stellarator research talks as yet about "power on the grid in 20 years" or so. They are already happy when they see some plasma being confined in their intricate vessels.

Stellarators and tokamaks have in common that they both use magnetic fields to confine the plasma in a torus-like device. The basic difference between them is how they cancel the particle drift due to the non-uniformity of the magnetic field, to make sure that the magnetic field lines form a closed surface inside the tube and do not end up in the wall of the tube, taking the plasma particles with them. To achieve this the magnetic field in both devices is helically twisted.

In tokamaks, a poloidal field is usually created by a toroidal electric current (a current travelling in the plasma around the torus, which also takes care of (part of) the plasma heating), in addition to a toroidal field generated by magnetic coils. In stellarators, only coils are used. There is no need for a current flowing in the plasma, which is *the fundamental difference between the tokamak and stellarator approach to nuclear fusion*. The absence of this current is a great advantage as it is a source of perturbations in the plasma and gives rise to instabilities. Since in a tokamak this current is also used to heat the plasma (ohmic heating) it implies that in stellarators other heating methods must be found.

In addition to toroidal-field coils, producing the toroidal field, the stellarator has one or several pairs of helical windings that complicate the design, but make it possible to create the required closed magnetic surfaces. Figure 14.1 shows the basic principle of a classical stellarator with 4 helical windings.

A second very important advantage of the stellarator over the tokamak is that it can operate in a steady-state regime, while tokamaks are almost by definition pulsed devices since the current through the central solenoid cannot be increased indefinitely. For a nuclear fusion power plant steady-state operation is obviously crucial. But, an advantage normally comes with a disadvantage, which for power reactors operating in a steady state is that the plasma needs to be refuelled continually and that impurity control and 'ash' removal (the residue from the burning plasma) are needed. For this they are equipped with a **divertor**. A divertor design was suggested as early as the 1950s by Spitzer. The magnetic field lines at the edge of the plasma were deliberately *diverted* into a separate chamber where the particles carried with

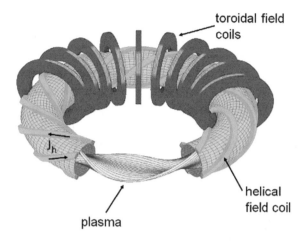

Fig. 14.1 Basic principle of a classical stellarator with helical coils. Note that the current flows in opposite directions in adjacent windings of these coils

them can interact with the wall and do no harm. As we have seen divertors are now also used in tokamaks.

Finally, the geometric parameters are also generally much different between tokamaks and stellarators. For tokamaks, the aspect ratio (the ratio R/a of major radius R and minor radius a, see Fig. 6.1) is usually in the range from 2.5 to 4, and even smaller for spherical tokamaks. To avoid certain resonances, stellarators are designed to have higher aspect ratios (5–12). A consequence is that the effective plasma volume in tokamaks is much larger than in stellarators.

Stellarators have great flexibility and different stellarator configurations (and combinations of them) have been developed. They go under various names, but no accepted usage of names has been established. The original figure-8 design is sometimes referred to as a **spatial stellarator**, while the name **classical stellarator** (Fig. 14.1) is used for a toroidal or racetrack-shaped design with separate helical coils to create the closed magnetic surfaces. The set of toroidally continuous helical windings with current flowing in opposite directions in adjacent windings provides the poloidal field. Because of the oppositely directed currents in adjacent windings there is no net toroidal field in such a configuration. A toroidal field must then be obtained from a separate set of coils (indicated in Fig. 14.1 as the toroidal-field coils). One of the advantages of such a setup is flexibility, as the helical and toroidal-field components can be varied independently. A disadvantage is that the interaction between the toroidal-field coils and the helical windings may cause large radial forces, posing serious problems for supporting the helical windings.

These designs were followed up by a design called a **torsatron** (known in Japan as **heliotron**). In a torsatron the currents in the helical field coils flow in the same direction (unidirectional). The 'basic' torsatron retains poloidal-field coils to cancel the vertical field produced by the helical coils, while the 'ultimate' torsatron design ('ultimate' meaning that a special winding for the helical field coils is selected) dispenses with the toroidal-field coils altogether, but requires the windings of the helical coils to be carefully specified. So this would essentially be the setup of Fig. 14.1 without the toroidal-field coils and with unidirectional helical field coils.

A further stellarator-type design is the **heliac** (helical axis stellarator), in which the magnetic axis (and the plasma) follow a twisted (helical) path to form a toroidal helix rather than a simple ring shape. The twisted plasma induces twist in the magnetic field lines to achieve drift cancellation.

The most important of the newer stellarator designs is, however, the **helias** (Helical-Axis or Helical Advanced Stellarator) configuration. It is the design of the most important stellarator currently in operation: Wendelstein 7-X. 'Wendel' is the German word for a helix or spiral, which explains the name. It is also a reference to the Wendelstein, which is the highest peak (1838 m) of the Wendelstein massif in the Bavarian Alps.

Wendelstein 7-X is located at the Max Planck Institute for Plasma Physics in Greifswald (IPP Greifswald) in the northeast of Germany, a branch institute of the Max Planck Institute for Plasma Physics in Garching near Munich (IPP Garching). The construction of Wendelstein 7-X took from 1991 to 2014 at a cost of slightly more than 1 billion euro and became operational at the end of 2015.

It is the successor of a number of earlier Wendelstein devices constructed from 1960 onwards, which makes Germany the most persistent player in the stellarator field. One of its predecessors gave a major boost to the stellarator line, as in 1980 it solved the heating problem caused by the absence of a plasma current. It was able to demonstrate for the first time the "pure" stellarator principle with a hot plasma, i.e., confinement without plasma current, showing that stellarators were back in business.

The Wendelstein 7-X helias configuration uses a single optimized modular coil set (i.e., a set made up of separate units consisting of a number of the strangely shaped coils, as depicted in Fig. 14.2, and some planar (flat) coils) designed to simultaneously achieve high plasma currents and good confinement of energetic particles. The planar coils are used to make fine adjustments

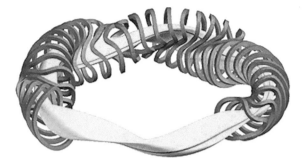

Fig. 14.2 Stellarator design with a single set of twisted external coils (like the one in Fig. 14.3), as used in Wendelstein 7-X, in addition to other coils. A series of non-planar magnet coils (blue) surrounds the plasma (yellow). A magnetic field line is highlighted in green on the yellow plasma surface

to the magnetic field. Figure 14.2 shows the schematic design of the device and Fig. 14.3 a non-planar coil as installed in it. Such a coil is about 3.5 m high.

The principal objective of Wendelstein 7-X is to investigate the suitability of this type of device for a power plant. It will test an optimised magnetic field for confining the plasma, with temperatures up to 100 million degrees and confinement times up to 30 min (much longer than for tokamaks). The field will be produced by its system of superconducting magnet coils. This is the technical core piece of the device.

It will be clear from Figs. 14.2 and 14.3 that the fields produced by these awkwardly twisted coils are very complex. The complexity of the stellarator magnetic fields makes them much more difficult to analyse than the fields in a tokamak. In this respect, however, stellarator design has hugely profited from dramatic improvements in computer power and three-dimensional modelling capabilities in the last few decades. This more than cancels out the complexity disadvantage of stellarators compared to tokamaks.

Wendelstein 7-X is the first large-scale, fully optimised stellarator and the world's largest device of its kind. It uses superconducting magnets, which implies that it is also a cryogenic device and has to be kept at the very low temperature of −270 °C (to be cooled to this temperature with liquid helium). Figure 14.4 shows the layout of the intricate coil system, consisting of 50 non-planar coils, 20 planar coils and five so-called trim coils. These external trim coils, of which four are shown in the figure, are used to fine-tune the shape of the plasma.

Around the coils we have the cabling, piping and central support ring, followed by the outer vessel with its diagnostic ports resulting in the bizarre

Fig. 14.3 Drawing of one of the non-planar superconducting coils as installed in the Wendelstein 7-X device

structure shown in Fig. 14.5, which shows little resemblance to any of Spitzer's original designs.

Plasma equilibrium and confinement in Wendelstein 7-X is expected to be of a quality comparable to that of a tokamak of the same size. With plasma discharges lasting up to 30 min, Wendelstein 7-X is to demonstrate the essential stellarator property, viz., continuous operation. Its primary purpose is to investigate whether it is suitable for extrapolation to a fusion power plant design. The future will tell whether that is the case. First plasma was produced at the end of 2015 and ignition conditions have not yet been reached.

The successor to Wendelstein 7-X, providing that the design proves suitable for a power plant, will be much larger and consequently much more expensive (comparable to the costs for ITER perhaps), most probably too expensive to be financed by a single country. The problem in this respect is that, if ITER fails, there will be little appetite for another global venture of

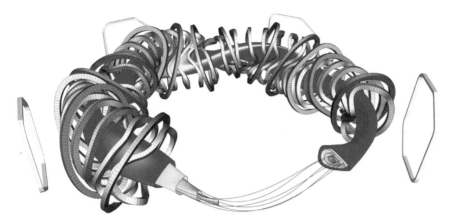

Fig. 14.4 Layout of the coil system of Wendelstein 7-X. Some nested magnetic surfaces are shown in different colours in this computer-aided design (CAD) picture, together with a magnetic field line that lies on the green surface. The coil sets that create the magnetic surfaces are also shown, planar coils in brown, non-planar coils in grey. Some coils are left out of the rendering, allowing for a view of the nested surfaces. Four out of the five external trim coils are shown in yellow. The fifth coil, which is not shown, would appear at the front of the picture

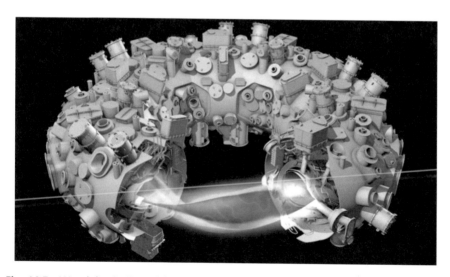

Fig. 14.5 Wendelstein 7-X with all its fittings and trimmings. A 16-m wide container enclosing all the magnetic coils and their helium cooling liquid, with 250 access ports

a similar scale and, while if ITER succeeds, the international fusion effort is likely to continue along the tokamak road, so it is hard to see any future for the stellarator as a power plant.

Revival of Magnetic Mirrors

As related very briefly in Chap. 6, in the early period of fusion research the mirror device was seen as an attractive method for confining hot plasmas. It was based on a straight cylinder, the simplest geometry for plasma confinement one can think of. The problem is of course that plasma particles can escape at both ends. Such leaks can be greatly reduced by forming two 'magnetic mirrors', which means that additional magnetic coils are installed to increase the magnetic field strength at the ends, to pinch the endpoints of the cylinder. The plasma particles are repelled by the stronger end magnetic fields (the mirror fields), containing them within a 'magnetic bottle'. However, good confinement could never be achieved in such devices and research on mirrors almost completely stopped in 1986 when the MFTF was promptly shut down after its completion, without ever having operated. The support for mirror research was terminated as the Department of Energy lent its ear to the tall stories of fusion power generation by the tokamak within 20–30 years.

However, some people did not want to give up and continued working on magnetic mirrors, especially in Russia and Japan. Two developments since the 1980s are the Gas Dynamic Trap (GDT), which operates at the Budker Institute of Nuclear Physics (BINP) in Akademgorodok, near Novosibirsk in Russia, and the Gamma-10 experiment in Tsukuba, Japan.

We will not discuss them here, but just mention that the current interest in such systems has been raised because the open magnetic field line configuration of mirror devices is particularly well suited for propulsion system applications, since they allow for the easy ejection of plasma that is used to produce thrust. An example of such an application is the Variable Specific Impulse Magnetoplasma Rocket (VASIMR), an electrothermal (magneto)plasma thruster under development at the Ad Astra Rocket Company in the US for possible use in spacecraft propulsion. In these engines, a neutral, inert propellant is ionized and heated using radio waves. The resulting plasma is then accelerated with magnetic fields to generate thrust.

Some Other Non-mainstream Magnetic Confinement Approaches

Non-mainstream approaches often combine aspects of magnetic and inertial confinement fusion. We will not discuss them in any detail, but just make a few remarks.

In any quest, even in the quest for fusion, Mother Nature sometimes is kind and seems to help. Certain plasma configurations exhibit properties known as "self-organization" where the plasma tends to converge to a certain optimum state; the plasma alters an externally applied magnetic field in a way that improves the confinement properties needed for fusion. Examples include approaches that go under the names of reversed-field pinch, spheromak, field-reversed configuration and magnetized target fusion, but tokamaks, too, show some features of plasma self-organisation, although they still require external control.

Self-organization is a generic process in nature, resulting in the spontaneous formation of ordered structures. In the universe it is common for plasma and magnetic fields to evolve together in a turbulent way, but then to rapidly relax to simple, self-organized structures. An example is magnetic reconnection, which happens for instance on the Sun. Solar flares erupt from the photosphere tangled and chaotic, but relax and straighten via magnetic reconnection with the release of huge quantities of energy. Such a self-organization principle forms the basis of several fusion approaches. One is the **reversed-field pinch** (RFP) which is a toroidal pinch, like the ones used in the early days; a torus with a strong poloidal field that can pinch the plasma (Chap. 6). The main idea of this version is to have the toroidal magnetic field, i.e., the magnetic field lines going around the torus the long way, run in one direction inside the plasma and in the opposite direction outside the plasma, giving rise to the term 'reversed field'. This field is then complemented by the poloidal field produced by a toroidal plasma current. Such a configuration can be sustained with comparatively lower fields than a tokamak configuration of similar power density. The main problem is that plasma confinement in the best reversed-field pinches is only about 1% as good as in the best tokamaks. This makes them lousy candidates for fusion reactors.

A **field-reversed configuration** (FRC) is an ultra-compact axisymmetric toroidal configuration in which a plasma is confined on closed magnetic field lines, like tokamaks and stellarators, but without a central hole. There is no central solenoid passing through the centre of the device. It has zero toroidal magnetic field everywhere and therefore no toroidal field coils. The plasma is confined solely by a poloidal field. The lack of a toroidal field means that this configuration has no twisted field lines and that it has a high *beta*. This makes it in principle attractive as a fusion reactor and well-suited for aneutronic fuels (no neutrons in the reaction products) because of the low magnetic field that is required. If the plasma current is strong enough, the magnetic field direction along the axis will be reversed. A compact toroidal plasma configuration may result, but field reversal alone is not enough to ensure that the field lines are closed.

These configurations were a major area of research in the 1960s and into the 1970s. The experiments so far have been small and had problems scaling up to practical fusion triple products. Interest returned in the 1990s and, as of 2019, FRC has been an active research area. A large number of experiments with field-reversed configurations have been and are still being carried out. The configuration is, however, highly unstable and must be carefully shaped to avoid instabilities. Private companies have emerged that study such configurations for electricity generation. We will come back to them in Chap. 15.

Spheromaks are FRC-like configurations with a finite toroidal magnetic field. These devices are sometimes also called magnetic vortices, magnetic smoke rings or plasmoids. The choice of the name spheromak, proposed in 1978, specifically stresses its relationship with the tokamak, because of the similar torus-shaped plasma that eventually forms. It has a low aspect ratio like the spherical tokamak and a short definition would be: a tight aspect ratio pinch with the toroidal field generated solely by plasma currents.

It is a toroidal plasma in a chamber without a hole in the middle. So there is no central solenoid and there cannot be any coils to generate a toroidal magnetic field. Thus, spheromaks manage to have a toroidal field (generated by the plasma currents) without having toroidal-field coils and without an ohmic transformer through the middle. They had some successes in the 1970s and 1980s, but by the late 1980s the tokamak had surpassed the confinement times of spheromaks by orders of magnitude. This was probably the reason why funding dried up. In the late 1990s research demonstrated that hotter spheromaks have better confinement times, which led to a second wave of spheromak machines.

The greatest difference between these three configurations (RFP, FRC and spheromak) and tokamaks and stellarators is that they all have much smaller or zero toroidal fields. A possible reactor based on these concepts would lead to a more compact device, reducing capital costs. The smaller toroidal field, however, results in poorer plasma performance. Transport losses are higher than for tokamaks and each configuration would be unstable without the presence of a perfectly conducting wall near the plasma surface. The clear advantage compared to the tokamak is that no high-field (superconducting) toroidal magnets are needed. The current drive problems for FRC, RFP and spheromak are also more severe than in tokamaks. Each of the alternative concepts improves one aspect of the tokamak concept but suffers with respect to others, and therefore does not obviously offer better prospects.

The idea of **cusp confinement**, so named because of the cusp shape of the curve in which opposing magnetic force lines meet, goes back to the early days of fusion. At the 1954 conference at Princeton, and especially as a result of Edward Teller's talk at that conference (see Chap. 6), it was concluded that any device with convex magnetic field lines towards the plasma would likely be unstable, resulting in kink instabilities and the like, and that a configuration with field lines curving away from the plasma should be inherently stable. Such a system would be a much better trap.

The basic geometry of a **biconic cusp** is shown in Fig. 14.6a, and a cross-section of this configuration in Fig. 14.6b. The cusp magnetic fields are formed by two circular coils with opposite currents. It is a mirror-like setup with the fundamental difference that the currents have the opposite direction, while in a mirror they are in the same direction. The magnetic fields are blocked from entering the high-pressure plasma by the plasma's internal magnetic field and the plasma is kept confined in the middle. The plasma presses against the outside cusped magnetic fields. This means that the magnetic pressure is equal to the plasma pressure and *beta* will automatically be 100%. An ideal situation.

At the centre there is a null point in the magnetic field (a zero-field point). The important advantage of the cusped configuration is its stability due to the favourable curvature of the external magnetic field towards the confined plasma system in the centre.

The chief disadvantage is the large rate of particle loss, as can be imagined, by diffusion along the magnetic field; particles squeeze out between the magnetic force lines where they meet at the cusps. This loss increases with temperature. Nevertheless, the confinement of a plasma in magnetic fields with such a geometry is an improvement over the simpler magnetic mirror discussed in Chap. 6.

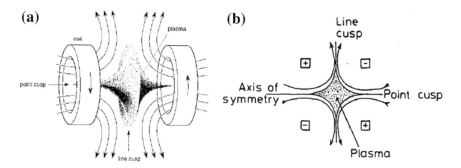

Fig. 14.6 **a** Cusped magnetic field, produced by two coils carrying currents in opposite directions. **b** Cross-section of the geometry shown in **a**

Papers on cusp confinement were already presented at the 2nd Geneva Conference in 1958 and it was fairly popular it seems in the 1970s. However, most cusped experiments failed and disappeared by 1980. Biconic cusps were recently revived by two private companies using similar geometries for designs of fusion reactors.

Inertial Electrostatic Fusion

Amasa Bishop in his 1958 book on *Project Sherwood*, the early American fusion effort, says about confinement by electric fields: "While not unequivocally ruled out, the use of electric fields for confinement does not appear to be feasible. The major deterring factor is that an electric field exerts oppositely directed forces on the electrons and positive ions of the plasma; if made to confine one component, the other would tend to escape. (…) Numerous proposals have been made for plasma confinement by electric fields, but none appears to have sufficient merit to warrant serious consideration."

He seems to be right, but in spite of this we will consider here its history and continuing fascination for a number of people. Most inertial electrostatic fusion devices directly accelerate their fuel to fusion conditions, thereby avoiding the energy losses seen during the longer heating stages of magnetic fusion devices.

Accelerated in an electric field, negatively charged electrons and positively charged ions in a plasma move in different directions and the field has to be arranged in some fashion so that the two species of particles remain close together. In most designs this is achieved by pulling the electrons or ions across a potential well, beyond which the potential drops and the particles continue to move due to their inertia. Fusion occurs in this lower-potential area when ions moving in different directions collide. The motion provided by the field creates the energy level needed for fusion, not random collisions with the rest of the fuel, and therefore the bulk of the plasma does not have to be hot and the system as a whole works at much lower temperatures than magnetic fusion devices.

A number of theoretical studies have pointed out that the inertial electrostatic fusion approach is subject to a number of energy loss mechanisms that are not present if the fuel is evenly heated, i.e., in thermal equilibrium. These loss mechanisms appear to be greater than the fusion rate in such devices, meaning they can never reach fusion breakeven and cannot be used

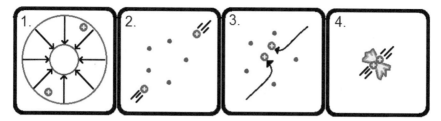

Fig. 14.7 Illustration of the basic mechanism of fusion in fusors. (1) The fusor contains two concentric wire cages. The cathode is inside the anode. (2) Positive ions are attracted to the inner cathode. They fall down the voltage drop. The electric field does work on the ions, heating them to fusion conditions. (3) The ions miss the inner cage. (4) The ions collide in the centre and may fuse

for power production. The assumptions made in these studies have been criticised and attempts are still being made by various (private) companies to develop inertial electrostatic fusion devices, but with scant success.

The best known and one of the simplest and earliest inertial electrostatic fusion devices is the **fusor**. Its basic mechanism is explained in Fig. 14.7.

It consists of two concentric metal-wire spherical grids. When the grids are charged to a high voltage, the fuel gas ionizes. The field between the two grids then accelerates the fuel inwards and heats the ions to fusion conditions. When passing the inner grid, the field drops and the ions continue inwards towards the centre. If they impact with another ion, they may undergo fusion. If they do not, they travel out of the reaction area and again into the charged area, where they are re-accelerated inwards. Fusors are popular with amateurs, because they are easy to construct, can regularly produce fusion, and are a practical way to study nuclear physics, but need energy to work.

They were invented in the 1950s by Philo Farnsworth (1906–1971), one of the relatively unsung heroes of the twentieth century, who was also the inventor of electronic television and apparently coined the name inertial electrostatic fusion. In his work on vacuum tubes for television he observed that electric charge would accumulate in certain regions of the tube. Today, this is known as the multipactor effect, invented by Farnsworth, and named as such by him. He filed a patent application for it in 1935. A particular variant of the multipactor featured two hemispherical electrodes that were placed near the perimeter of a glass sphere, and another electrode–a wire grid–was placed at the centre. When Farnsworth operated his device, he observed a tiny, star-like anomaly suspended within the inner grid at the centre of the tube. Increasing the power level, the point of light became brighter still and, even more impressively, it never touched the walls of the tube itself. Apparently, without realizing it at the time, Farnsworth had created the first fusor. In

1962, he filed a patent on a design using a positive inner cage to concentrate plasma, in order to achieve nuclear fusion.

In 1959 it was already concluded on theoretical grounds that "[a]lthough it is of doubtful utility as a thermonuclear reactor, it may be possible to produce in this way small regions of thermonuclear plasma for study. The device appears to be unstable at economic densities."

Indeed, no fusor has come close to producing a significant amount of fusion power. A fusor can be built at home and produces fusion reactions, but hardly any power and must be connected to a power supply to work. On the other hand, "the fusor just looks totally cool. An eerie purple-blue glow emanates from the reactor, and a really well-made fusor can produce a mesmerizing phenomenon called a 'star in a jar'." They can however be dangerous if no proper care is taken because they require high voltages and can produce harmful radiation (neutrons and X-rays).

In the next chapter we will pay attention to a few private companies that have recently revived some of the ideas considered in this chapter as well as others.

15

Privately Funded Research

The current situation in nuclear fusion research, contrary to its first decades, is that most non-mainstream fusion approaches, both as regards devices and alternative fuels, are privately funded. Virtually all public money (globally about $2 billion per year) goes into tokamaks, especially into the cash-hungry ITER and other toroidal devices that can show themselves to have some 'ITER-relevance'. Consequently, the opportunities to develop new ideas and start alternative fusion projects at national and/or university laboratories are dwindling. In view of the fact that it apparently seems to take forever for fusion to become a reality, it is not surprising that governments are not rushing to fund all kinds of wild, or even sensible ideas or projects in this field. So, venture capital can play a useful role here at no extra cost to the taxpayer. For venture capitalists it is just a wager, a little money invested in a project that has little chance of working, but promises a huge payoff if it does. The chance of success should not be zero, though, else it would just be foolish to invest.

The business model of private companies working in the fusion field is to do (good) research with less overhead than universities or national labs and with the bold intention of producing commercial fusion power or something else of use within a couple of decades at the most, while drawing on contributions from venture capitalists and at the same time competing for the same government grants, e.g., from the US Department of Energy, taking advantage of money earmarked for small businesses, for instance. Quite a few of these companies are spin-offs from universities or national labs, using

L. J. Reinders, *Sun in a Bottle?... Pie in the Sky!*, https://doi.org/10.1007/978-3-030-74734-3_15

ideas that have come to fruition at these labs and now need funding to test them in the marketplace. Innovative technologies can only have an impact on society if they emerge from the lab and enter the marketplace. Governments and universities fund basic research, but stop well short of the level of development required for commercialization. Industry and investors are reluctant to pick up that technology when the payoff is longer than a few years and when substantial technical and economic factors remain unresolved. This is what MIT calls the "valley of death". Having ideas that are technically sound is not enough—there must be a viable pathway to widespread, profitable deployment in order to succeed. Technologies, like fusion energy, that require substantial investments and take many years to reach fruition face a particularly deep and broad valley. It is claimed that investments in fusion by the US government have not been in proportion to its potential to meet the threat posed by climate change, even in its funding of basic research on fusion science and technology. That claim is not correct, if only because the potential of fusion to meet this threat is mostly fake, in spite of all the funding that has been sunk into fusion research in the last 70 years. That funding has gone on for far too long and it is now time for private capital to waste some of its money on this impossible venture. On a path determined by current levels of public funding, in the US and elsewhere, fusion energy is unlikely to come on stream before this century is over. The new model is to employ private companies stuffed with private capital and academics full of expertise gained at government and university labs—reaching across the valley of death. This new model seeks a much more rapid development path, claiming even that it could develop fusion power in time to address the growing threat that humanity faces from climate change.

Most of these companies have sprung up in the fusion landscape in North America, especially in the US, where the surroundings of Seattle in Washington State are especially well represented. The reason is probably that funding for fusion research at universities and national labs has been squeezed more in the US than in Europe. There are some private ventures in the UK (e.g., Tokamak Energy which we met in Chap. 13) and there are one or two on the continent. One Australian and one Chinese company complete the picture.

As is common in fusion land, newspapers have already been raising overblown expectations on what these companies will achieve. 'Limitless' and 'inexhaustible' in respect of the promised energy are favourite epithets in this respect. The Guardian Weekly newspaper of 16 March 2018, for instance, announced 'fusion power to be on grid in 15 years', when describing the collaboration of scientists at MIT with the private company Commonwealth

Fusion Systems performing experiments with a tokamak, which is going to make its benevolent appearance just 'in time to combat climate change'. What a relief that we no longer have to worry about that. There can be no doubt though that the best contribution from these endeavours to combating climate change would be to immediately stop the experiments, since their carbon footprint is just gigantic. There are stories of this kind for almost all the alternative approaches tried by private companies. They were all tried in the past and judged to be lacking promise and yet suddenly there are now solutions all over the place. These start-ups all want fusion quicker and cheaper (who wouldn't?) and go for small devices, much smaller than the ITER colossus, which by the way is not even a power plant. Spherical tokamaks and/or high-temperature superconducting magnets among other things must do the trick. All have several features in common. The first is that none has ever produced a single watt of power from nuclear fusion reactions, although they still claim to have "power on the grid" within 10–20 years or even less; and secondly, they all claim to be ahead of each other and of all other fusion energy technologies including the ones that actually have produced fusion reactions and/or power, albeit very little. The peculiar thing about all this is that the press is just printing these silly claims without asking any further questions. History is repeating itself where their websites dangle in front of our eyes the prospect of working fusion power plants within an impossibly short term, while at the same time casually mentioning a solution to climate change. This can never hurt as it may entice panicky governments into loosening the purse strings and dole out some taxpayers' money as a bonus. The phrases "virtually unlimited energy" or "inexhaustible source of energy" are also effective in this respect, but a bit stale as they have been in heavy use for close to seventy years now. And all this while nothing fundamental has changed, and there is no reason whatsoever to think that they will succeed where others have failed for so long. Would the plasma suddenly stop its unruly behaviour because of a relatively small change in the geometry of the torus (more spherical) and start fusing like hell in an orderly fashion and/or click into place at the command of a set of more powerful magnets? Some say that advances in supercomputing and complex modelling are the breakthrough that will do the trick. Modelling plasma behaviour and using supercomputers to grind through the hideously complex models can indeed help to bring long overdue knowledge on plasmas within reach and can speed up the design process, but it cannot do much else. It doesn't change the fundamentals. It is very unlikely in my view that it will help much. An ELM or two will most probably be enough to teach the necessary lesson.

But whatever the motivation, the fact is that private capital invested in this kind of project has soared in the last decade, with over $1bn of private venture capital poured into fusion projects in the United States alone. This money is also encroaching on mainstream fusion projects, where the tendency to divert all public funding into the mainstream tokamak approach seems to be changing somewhat, the major example being the UK government investing quite heavily in a spherical tokamak program, called STEP. In Chap. 13 we already considered private investment in spherical tokamaks, especially those made by Tokamak Energy in Britain. Although Tokamak Energy with its spherical device can still be considered non-mainstream, albeit just marginally, the private company Commonwealth Fusion Systems, founded in 2017 as a spin-off from MIT, uses a conventional tokamak design with high-temperature superconductors to take the fast route to fusion. Here, in this chapter, we will pay a little more attention first to this company and then briefly to the various non-mainstream approaches to nuclear fusion embarked on by private companies with alternative devices, some of which have already been mentioned in the previous chapter.

Commonwealth Fusion Systems (CFS)

CFS is a spinoff from MIT and headed by former students and scientists of the MIT Plasma Science and Fusion Center (PFSC). Like MIT it is based in Cambridge, Massachusetts, and it funds $30 million of research at MIT. MIT's motivation to follow the route of collaboration with private capital has been formulated in its SPARC brochure and has to do with "reaching across the valley of death" as explained above. It is the private company Common-wealth Fusion Systems that collects money (about $200 million so far) from venture capitalists who apparently value its commercial potential, and uses this money to pay for research carried out at MIT. Over the long term, CFS seeks to commercialize fusion and lead the world's fusion energy industry. So, a grand vision indeed. For the time being, MIT and CFS will collaborate on the research and development of high-temperature superconducting (HTS) magnets and on the SPARC experiment.

Its timeline is to produce net energy by 2025 and generate power on the electrical grid by 2036. It is unimaginable that they can adhere to that time-line, considering that currently, late in 2020, they are still developing and testing the HTS magnets and no plasma has been produced yet. Even a coal plant would have difficulty meeting that deadline.

In collaboration with MIT, the SPARC demonstration plant is currently being built. It is claimed to be the world's first fusion device that produces plasmas which generate more energy than they consume, becoming the first net-energy fusion machine. The SPARC goal is to exceed $Q = 2$, in a significantly smaller device than JET. Whether that means net energy depends on what its energy consumption will be and what is included on the energy balance sheet. SPARC will have a plasma volume of 15 m^3, the same as a mid-sized fusion experiment—similar to many machines already in operation. It is the new HTS magnet technology that must do the trick. These magnets can be run at much higher field strengths, roughly doubling the magnetic force on the fuel. In this respect CFS follows the same path as Tokamak Energy. Like the Tokamak Energy machines, the CFS machines are high-field, high-density devices. There is no experience with such devices and it would be very surprising indeed if the plasma did not come up with some surprises.

SPARC is claimed to be about the minimum size experiment that could make a net-energy plasma using this magnet technology. While the SPARC magnets will be superconducting and run in steady state, the plasma will be pulsed—lasting about 10 s—to simplify many aspects of the device.

CFS estimates that its burning-plasma device, SPARC, will cost around $400 million, which is just small change compared to the more than $20 billion construction costs of ITER. But such a comparison is of course not fair, as ITER intends to achieve Q > 10 and is based on old technology, in particular ordinary instead of high-temperature superconductors. If ITER had to be designed now, it would be a completely different device.

The development of SPARC relies heavily on data from decades of research on dozens of experiments around the world, including data compilation and analysis during the ITER design stage. This gives CFS a major advantage in the early development stages, making fast progress possible so long as no new ground has to be covered. At a certain point it will enter uncharted territory and things may begin to look less rosy. It would be the first nuclear fusion device that does not run into unexpected complications.

Following the SPARC demonstration which is scheduled to start in 2021, but unlikely to happen, CFS will construct the "world's first fusion power plant", based on the ARC tokamak concept. ARC (short for affordable, robust, compact) is a theoretical design for a compact fusion reactor developed by the MIT PFSC and has a conventional advanced tokamak layout. It is not a spherical tokamak, although the design is compact. The advanced tokamak aims to increase tokamak performance by using active control of the cross-section shape (D-shape) and simultaneously optimizes both *beta* and confinement in a manner consistent with steady-state operation. The ARC design also uses HTS magnets, of course.

ARC is a tokamak with a physical size slightly larger than any of the big tokamaks (JET, TFTR, JT-60), but with a magnetic field that is two to three times larger. It is designed to produce ~200–250 MW in electricity (525 MW of thermal fusion power). ARC is significantly smaller in size and thermal output than most current reactor designs, which typically seek to generate ~1 GW in electricity.

The HTS technology is key to CFS's approach. The claim is that "advances in superconducting magnets have put fusion energy potentially within reach", so without these HTS no fusion, and with HTS only potentially so. The most important advantage of these HTS superconductors is that they can sustain much stronger magnetic fields. The confinement time for a particle in a plasma varies with the square of the size of the machine and the fourth power of the magnetic field, so doubling the field offers a four times larger performance for the same size of machine. The smaller size reduces construction costs, although this is offset to some degree by the expense of the HTS magnets.

The second important point is that conventional LTS magnets as used in ITER would need thick shielding (of about 1 m) to prevent the high-energy neutrons produced in the fusion reactions from heating the superconductors. HTS magnets potentially provide a solution for this. They become superconducting at temperatures of around 90–100 °K (−183 to −173 °C), so quite a difference, although there is still a fair amount of shielding needed, which is the reason why ARC cannot utilise an ST design. But it means that liquid neon, hydrogen or even nitrogen can be used for cooling rather than liquid helium, making the cryogenics simpler and cheaper.

The promises made by CFS are sky high, a foolish attitude that is common in fusion land (see Tokamak Energy in Chap. 13). A little more modesty would probably help them to go further.

Privately Funded Non-mainstream Approaches

One of the surprising aspects of the various private companies working on fusion is that they immediately go for *commercial* energy production, as if (net) energy production by nuclear fusion as such is a well-established process that has been well and truly proved. It is as if, within the last twenty years or so, quite a number of promising, fast routes to *commercial* energy production from fusion have suddenly sprung up where before there were none, and where all the efforts of the last 70 years were essentially in vain. It just doesn't make sense.

Worldwide there are now a few dozen such companies, taking on the challenge in which public funded endeavours have so miserably failed. Some are very young, others are older, but most were founded in this century. Some disappeared after a few years. The world is awash with cheap money that has to be spent somewhere, so why not redo some of the failures of the past, to remind us that history had a reason for unfolding as it did. They are all very bold in their claims, outrageously so, insofar as they reveal anything about their intentions. They have in common that they want to revolutionise things by promising quick, abundant, clean, inexhaustible energy at a fraction of the cost spent on tokamaks, and in that they will probably all fail. Reading the blurb on their websites one after the other, it is quite painful to see this mantra of clean, safe and abundant energy being repeated all the time.

It should be said though that not all companies have the goal of creating energy to sell on the grid, but want to use nuclear fusion for other purposes such as rocket propulsion. Examples are Ad Astra Rocket Company, mentioned above, and **Princeton Satellite Systems**, which licenses patents from Princeton University for space propulsion applications.

In the following we will just mention some of the most high-profile companies without going into detail about what they are actually doing. If the reader wishes to know more about any of them, they should consult the appropriate website. A list of websites is provided at the back of this book.

There are companies (**General Fusion, Helion Energy, Compact Fusion Systems** and **MIFTI**) that work on some form of magneto-inertial or magnetized target fusion (MTF), which combines magnetic and inertial confinement.

In particular, **General Fusion** has been around for quite some time now. It is a fairly large company with more than 70 employees and has managed to attract funding from various venture capitalists, among whom Amazon founder Jeff Bezos through his investment vehicle Bezos Expeditions. The fusion device under development at General Fusion is rather original; it has no vacuum chamber but uses a sphere (approximately 3 m in diameter) filled with a molten lead-lithium mixture. By rotating the metal mixture, a vortex (a form of turbulent flow revolving around a straight or curved axis) is created at the centre of the sphere. A pulse of magnetically-confined deuterium–tritium plasma fuel is then injected into the vortex. Around the sphere, an array of pistons drive a pressure wave into the centre of the sphere, compressing the plasma to fusion conditions, releasing energy in the form of fast neutrons.

This process is then repeated, while the heat from the reaction is captured in the liquid metal and used to generate electricity via a steam turbine. Since 2017 General Fusion has been developing subsystems for use in a prototype to be built in three to five years. So, 2022 should be a crucial year for them.

Helion Energy wants to build its Fusion Engine on deuterium-helium fusion, to avoid the troublesome high-energy neutrons produced in D-T fusion. Deuterium and helium are heated until a plasma is formed, after which pulsed magnetic fields accelerate the plasma into the burn chamber at over 1 million miles per hour. A strong magnetic field compresses the plasma to fusion pressure and temperature. The deuterium and helium nuclei fuse, releasing charged particles (helium nuclei). Their energy is directly converted into electricity and the helium is used for starting up the next cycle. The hope is to produce 50 MW of power in modules the size of a shipping container, with the aim of having a commercial plant operational in six years (from 2020).

The very young **Compact Fusion Systems** and **MIFTI** work on similar ideas.

A number of companies seeks to revive Philo Farnsworth's inertial electrostatic fusion, in combination with cusp confinement and/or other methods. **Progressive Fusion Solutions, Fusion One Corporation, Lockheed Martin, EMC2, Horne Technologies** and **Convergent Scientific Inc** are among them.

EMC2 (Energy/Matter Conversion Corporation), which worked on a device that combines cusp confinement with inertial electrostatic fusion, called the polywell, seems to be dormant now due to a lack of funds.

Progressive Fusion Solutions, a very young company from Vancouver (Canada), has the fantastic ambition to develop a car-sized fusion energy generator to replace fossil fuel power plants and end climate change. Why not? You have to think big in this business. Apart from some general statements about "implementing fusion energy into society and ending climate change" the company has not yet anything to show for.

Convergent Scientific Inc. or CSI was founded in 2010. It developed a variation of the polywell approach to fusion, capable of producing extremely energetic non-neutral plasmas with much lower relative input energy requirements. The company seems to have gone under.

Started in 2008, **Horne Technologies,** developed the Horne Hybrid Reactor (HHR), which they claim is the world's first continuously operating, superconducting, high-*beta* fusion research device. In spite of its momentous claim it has attracted surprisingly little attention, much less in any case than its closest relative Lockheed Martin's Compact Fusion Reactor.

Fusion One Corporation was in business from 2015 to 2017. The company developed a magneto-electrostatic reactor, partly based on the polywell, with some novel additions to stem particle losses through the magnetic cusps. With these additions, a pathway to commercial electricity generation no longer seemed feasible.

That leaves **Lockheed Martin**, one of the largest companies in the aerospace, defence, security, and technologies industry. As of December 2017 it employed approximately 100,000 people worldwide, and is the world's largest defence contractor with revenues of more than $50 billion (2018). It has ample experience with research into nuclear fusion as the former owner and operator of Sandia National Laboratories from 1993 to 2017.

Lockheed Martin Skunk Works intends to build the Lockheed Martin Compact Fusion Reactor (CFR), a high-*beta* compact reactor to be achieved by combining cusp confinement and magnetic mirrors to confine the plasma.

On the website of its compact fusion endeavour Lockheed does not reveal much, only that it intends to build a "reactor small enough to fit on a truck [to] provide enough power for a small city of up to 100,000 people." It is in general very secretive about its efforts and apart from a poster at an APS conference in 2016 nothing substantial has been published. The project started in 2010 and in 2014 Lockheed Martin announced a plan to "build and test a compact fusion reactor in less than a year with a prototype to follow within five years." This period has been over for some time now.

Halfway through 2019 Lockheed announced that it was in the process of constructing its newest experimental reactor, known as the T5, and that, despite slower than expected progress, it remains confident that "the project can produce practical results, which would completely transform how power gets generated for both military and civilian purposes."

A third group of companies (**LPPFusion, HB11 Energy, TAE Technologies**) is trying to find the holy grail of fusion in alternative fuels, especially proton-boron plasmas. As already mentioned in Chap. 3, fusion of boron with hydrogen (p-^{11}B fusion) is extremely difficult, five orders of magnitude more difficult than the easiest D-T fusion, at a temperature of about 100 keV (1 billion degrees). Proton-boron fusion is a factor of 100,000 less efficient, which is the reason why this fusion option is usually excluded. This can be changed by igniting the plasma under conditions of local non-thermal equilibrium, which is the method that all these companies use, but each in its own way.

TAE Technologies (formerly Tri Alpha Energy), founded in 1998 and based in Foothill Ranch (California) claims to be the world's largest private fusion reactor company. It is indeed one of the larger players in fusion in the

US, but as a company it is dwarfed by Lockheed Martin or General Atomics, for which nuclear fusion is just a side issue. TAE grew out of the physics department at the University of California, Irvine. With its $600 million (others say $800 million) in funding from private investors, it has received a big chunk of the total private venture capital put up in the US. One of its investors was Microsoft co-founder Paul Allen, who died in 2018. His former colleague Bill Gates apparently invests in one of TAE's competitors, Commonwealth Fusion Systems, but it seems that Gates, with good reason, has more confidence in power from nuclear fission, judging by his company TerraPower. TAE is currently testing a full reactor and raising funds to build a next-generation device.

In March 2019 TAE reported a fusion yield of 40 microjoules, the only company apart from LPPFusion to do so, but the yield cannot have originated from proton-boron fusion.

LPPFusion's mission is the usual one "to provide environmentally safe, clean, cheap and unlimited energy for everyone," in their case through the development of Focus Fusion technology, based on their so-called Dense Plasma Focus device and hydrogen-boron fuel.

In 2016 an ion temperature of 2.4 billion degrees was achieved in the FF-1 experimental device, which also gave a fusion yield of 1/8 J. The fusion yield is still tiny, but the temperature is adequate for proton-boron fusion (although no proton-boron fuel was used). Experiments with hydrogen-boron fuel were supposed to start in 2019, but it did not happen. Troublesome rapid oscillations in the current caused delays. They are now hurrying up, I suppose, to meet their schedule of showing net energy production in 2020 and present a first prototype 5 MW generator in 2023.

The company **HB11 Energy**, a spin-off from the University of New South Wales in Sydney (Australia), makes very big claims indeed for laser driven non-thermal proton-boron fusion. It is a type of inertial confinement fusion, but with proton-boron fuel. The claims on their website beat all others and are simply amazing, including that HB11 Laser Boron Fusion Reactors, so far only existing on paper, will produce electricity at a quarter of the price of electricity from existing coal fired power stations and are much cheaper to build; the technology already exists and the first prototype could be built in just 5–10 years; only modest investment is required for further research to clarify some of the scientific methods and engineering requirements; the reactors can be built anywhere and need no water as they produce electricity directly, so there is no need for steam turbines; they can be built to scale, meaning that smaller communities can build smaller reactors, while larger reactors can be built for large scale urban and industrial usage; they can be used on ships and submarines, for industry and manufacturing, isolated communities or inter-connected to major power grids. In short, an almost free lunch for everybody, where Laser Boron Fusion Reactors offer an inexpensive, clean, long term solution, with no carbon emissions, no radioactive waste and a very small environmental footprint.

ZAP Energy is a spin-off from the University of Washington and Lawrence Livermore National Laboratory. Its website does not reveal much, only that magnetic coils are too expensive, inertial confinement is too inef-ficient, and conventional Z-pinches are too fleeting, so the most promising path is sheared-flow-stabilised Z-pinch, meaning that its technology stabilizes plasma using sheared flows—that is, with layers of plasma flowing at different velocities at different radii—rather than magnetic fields.

In 2015, ZAP Energy was awarded a grant from ARPA-E's ALPHA program and it exceeded all of its aggressive program milestones. What these are, however, is not explained. It sees its reactor as the least expensive, most compact, most scalable solution with the shortest path to commercially viable fusion. Based on progress to this point, it is on track to reach $Q = 1$ energy breakeven plasma conditions.

Another American company is **AGNI**, based in Washington State, a start-up pursuing an unconventional version of scaled-down fusion in which a beam of high-energy deuterium atoms is fired at a target of lithium and tritium. In their own words they are expected to achieve: 16 million times the efficiency of coal, 10 times the efficiency of fission, no waste and zero emissions. Isn't it fabulous?

The company **CTFusion** is trying to further develop the concept of imposed-dynamo current drive (IDCD) and the dynomak, a power-generating spheromak reactor. Halfway through 2019, CTFusion was awarded several million dollars from ARPA-E to develop the dynomak concept. It has enough similarities with the tokamak that much of the theory, codes and expertise of the tokamak can be applied.

Type One Energy is a University of Wisconsin-Madison spin-off, applying innovations in additive manufacturing (3D printing), analytical theory, and high-field HTS magnets to drive the stellarator fusion concept towards commercialization as a compact and cost-effective power plant. It has laid out a 15-year path to a pilot power plant, named the Spitzer-1 stellarator, which will have 800–1200 MWe power generating capacity, is about twice the size of Wendelstein 7-X, and should be ready around 2035. Type One Energy does not intend to use tritium as a primary fuel, but to utilize the catalysed D-D fusion cycle by recycling the tritium produced in one of the two D-D fusion reactions.

In spite of being at the very beginning of its 15-year development path and without having yet constructed any power-generating stellarator, Type One Energy already sees massive revenues "generated through the mass production and sale of stellarator power units produced from Type One gigafactories fabricating a high fraction of pre-assembled components to expedite construction for a 2-year on-site install target." They see themselves having a 15% market share of energy-generating capacity in 2050 by capturing 56% of new energy consumption. An incredible feat of wishful thinking.

Renaissance Fusion is one of the very few fusion start-ups in continental Europe. It was founded in 2019 and is based in Granada, Spain. Renaissance Fusion aims to develop subsystems for stellarators and tokamaks, one of the few companies that see prospects in stellarators.

Conclusion

Most of the companies considered here have in common that they promise a quick fix to the difficulties that have beset nuclear fusion research now for close to seventy years. The surprising thing is that they apparently do this with all the confidence in the world and that the venture capitalists are easily convinced of the merits of all the proposals and overblown expectations.

All this private activity has also had its effect on the public sector. In general, nuclear fusion is characterised by the fear of losing out. Imagine, however unlikely this may be, that a country or company cracks the fusion nut in the next decade or so. Then we, of course, also want to benefit. This is undoubtedly one of the motives for the British government to start the STEP program, and for the US and India, and perhaps other nations too, to continue to participate reluctantly, but on the cheap in the ITER program.

The private efforts of the last few years have also convinced the US Department of Energy that the private sector has important contributions to make in the quest for fusion energy. Why they think so is not clear as the private sector has so far not contributed anything of any significance. In 2019 the INFUSE program (Innovation Network for Fusion Energy) was announced, an effort to pair national laboratories with private companies. To apply for a grant a private company must partner with a national laboratory. The funding provided is rather low compared to the amounts put in by venture investment funds. However, the INFUSE program will allow companies to get major experiments done at national laboratories; big experiments on large scale equipment. The program does not provide funding directly to the private companies, but instead provides support to the partnering DOE laboratories to enable them to collaborate with their industrial partners, which include Commonwealth Fusion Systems, TAE Technologies and various companies that have not been mentioned here.

16

Criticism of the Fusion Enterprise

Since the dream of abundant and cheap energy associated with nuclear fusion is slow in coming, many have wondered whether it is worth all this effort and money. On the Internet the hostile comments about the waste of money going into nuclear fusion research are almost unlimited. Books have been written to belittle the fusion effort and cast doubt on the sincerity and integrity of the scientists involved. Such criticism is certainly not without ground. Although it probably goes too far to accuse fusion scientists of outright dishonesty in the pursuit of their dreams, some deception and cheating is certainly going on. When we read some of the careless statements by people calling themselves scientists, of which we have seen a few examples scattered throughout this book (see, e.g., the discussion about the term 'breakeven' in Chap. 9) and of which we will see more examples below, we may well wonder whether they know how science is supposed to work. Some see it as an exercise in publicity, it seems, with the simple goal of raking in as much money as possible to be able to continue on this endless road. If a little dishonesty is needed to achieve this, so be it; once the splendid dawn of cheap and abundant energy is upon us, all will be forgotten. This dawn, however, refuses to break and it can indeed be said without hesitation that the result of the huge effort spent on fusion research in the last 70 or so years has so far been deeply disappointing, to the extent that no one can be blamed for no longer believing in the promise ever being fulfilled.

The argument often used by proponents of nuclear fusion is that we do not have a choice; that fusion is necessary not only to combat climate

L. J. Reinders, *Sun in a Bottle?... Pie in the Sky!*, https://doi.org/10.1007/978-3-030-74734-3_16

change and provide a carbon-free energy source, but also to be able to meet the global demand for energy. The number of people on Earth is soon to reach a staggering 10 billion who all want to use electricity in ever greater amounts. Burning fossil fuels is no option. Not only will there not be enough in the long run, but the climate change consequences are becoming ever more pressing. Nuclear fission energy has been banned in many countries because of its waste problems and the risk of man-made (Chernobyl) and natural (Fukushima) disasters. In the long term the only resources capable of producing energy on a massive scale will be solar and wind in one form or another and fusion. Solar and wind energy need breakthroughs in production and storage before they can be deployed on the scale needed, which leaves nuclear fusion energy as the only alternative to be pursued. The pros, if it works, are obvious: abundant, high-energy density fuel, producing per gram of fuel about 10 million times more energy than gasoline or coal, no greenhouse gases, safe, minimal "afterheat", no nuclear meltdown possible, small residual radioactivity with short-lived and immobile products, minimal proliferation risks, minimal land and water use, no energy storage issue and no seasonal, diurnal or regional variation. Some of these advantages are undoubtedly real, but not quite as real as their proponents would have us believe, as we will see in this and the following chapters. The cons are only high capital costs and, most importantly, the fact that it does not work and probably will never work. This latter fact is of course lethal. From the description of the fusion effort so far in this book, it should be clear that an awful lot has still to be done to make it work. Success is in no way guaranteed, very unlikely even, and will surely not be achieved within the foreseeable future. No sooner than around 2050 could there be any certainty about whether or not large-scale energy generation from fusion is *in principle* possible or should be abandoned and left to the stars. And even if it works and power plants can be built, they will certainly come too late and the cost will probably be prohibitive, not to speak of the engineering problems and complexity of a fusion reactor. Power companies will be very reluctant to run them. They may be forced to take them on in order to survive, as other power plants will no longer be tolerated. It may even have to come to a situation in which there is no longer any place for private energy generating companies and (inter)governmental enterprises will have to take over power generation.

In any case, the technical and scientific challenges faced by fusion are impossible to overcome in the short time the world has available for making the switch to carbon-free power generation. Advances in renewable energy (and its storage) will probably manage to solve problems with future energy needs faster and at lower cost than fusion will ever be able to do, if at all.

Fusion will take far too long to reduce greenhouse gas emissions to avoid unacceptable global warming. Energy from nuclear fusion in any amount that could make a difference cannot be expected in this century. If all goes well, and that is indeed a big if, a first experimental power plant may be available at staggering cost at the very end of this century. We will discuss this in more detail later in this chapter.

First, I will consider the cost factor of fusion research, followed by the rosy pictures painted by scientists and the media about fusion's prospects. The rest of the chapter will be devoted to some of the (serious) criticism made of the fusion enterprise as a whole, of its unachievable goals, and of specific projects, such as ITER. The criticism comes mainly from (former) scientists. There are very few active fusion scientists who vent their critical observations, but a growing number of retired fusion scientists are starting to speak out against the fusion folly. It seems that as long as fusion pays for their mortgage and the schooling of their kids, they behave as ardent supporters of fusion and refrain from speaking their mind. Only in retirement can they forgo the pretence and actually speak out.

Cost

A much-heard point of criticism has to do with cost. That criticism does not wash. In actual fact, the amounts spent on fusion, currently annually just a couple of billion of dollars for all countries taken together, are not that large when compared with other relevant pursuits of mankind. It is of course a large amount of money and, if well spent (on trying to understand the science behind fusion for instance), could do a lot of good, but compared to the amounts sunk into energy research and exploration of fossil fuels, they are actually fairly small, not to speak of the costs due to oil spills and other disasters with ordinary fossil-based energy, or the sums spent in US oil subsidies. US tax breaks for oil and gas companies still amount to $4 billion a year. The BP oil spill in the Gulf of Mexico in 2010 is estimated to have cost about $40 billion and dwarfs the entire fusion effort, while the price of the Chernobyl and Fukushima disasters is indeed unfathomable. Nobody has any idea how high that price eventually will be or how to calculate it, but it will certainly be on the order of trillions of dollars.

Moreover, other renewable energy sources are not particularly cheap either. In Germany, for instance, subsidies for solar and wind energy have amounted to on average €25 billion per year in recent years. As can be seen from Fig. 16.1, the total amount spent on fusion energy in the US, since the fusion

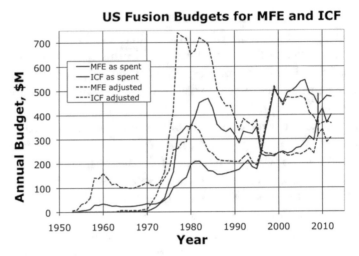

Fig. 16.1 US fusion budgets from 1950 to 2012. MFE stands for Magnetic Fusion Effort, ICF for Inertial Confinement Fusion. Adjusted means adjusted for inflation

programme began in 1953 until 2012 is $24.1 billion dollars (adjusted for inflation to $30.4 billion). That is an average of $400 million a year, adjusted to $515 million per year. It includes the expenditure on inertial confinement fusion, as well as on tokamaks and alternatives. The total 2016 US fusion budget was $951 million; by 2019 it had gone down to a mere $564 million and for 2020, thanks to action by Congress, it went up again to $671 million. As a comparison, NASA's budget in 2016 was eight billion.

As the figure also shows, US spending has been rather erratic, reflecting policy changes in the US over the years. It is argued that these policy changes, often caused by changing priorities each time a new administration took office, have regularly stymied attempts by the fusion community to capitalize on fusion's scientific successes and promises in the past. This argument is rather shaky in my view, but it is obvious that as it stands at present the US fusion budget is not on a track that will ever lead to a US demonstration plant.

Painting Rosy Pictures

It is common practice in the fusion business to paint rosy, if not dishonest, prospects. For instance, in the abstract to a 2005 "Status report on fusion research" published in a *scientific* journal by the International Fusion Research Council, an august body created by the IAEA in 1971 to advise the agency on its activities in the field of nuclear fusion, it is stated: "Fusion is, today, one of the most promising of all alternative energy sources because of the vast reserves of fuel, potentially lasting several thousands of years and the possibility of a relatively 'clean' form of energy, as required for use in concentrated urban industrial settings, with minimal long term environmental implications. The last decade and a half have seen unprecedented advances in controlled fusion experiments with the discovery of new regimes of operations in experiments, production of 16 MW of fusion power and operations close to and above the so-called 'breakeven' conditions." This is on the verge of outright lying, and in any case grossly misleading and unscientific, as fusion in its current state is in no way promising and the advances of the last decade or so may have been unprecedented (in the field of nuclear fusion not much is needed to be unprecedented, as there simply is no precedent), but certainly not promising or even encouraging. The fact that there are vast reserves of fuel is completely irrelevant if this fuel cannot be made to 'burn'! If anything, the whole fusion enterprise looks hopeless. Breakeven conditions are still very far away. The use of the word 'breakeven' in the quote above is aptly (but unscientifically) qualified by 'so-called', as 'breakeven' in nuclear fusion, as we have seen in Chap. 9, is a completely different concept from what a normal sensible person would mean by it, and is used by fusion proponents in a very misleading way, and probably with the intent to mislead. It is in any case not true that there have been operations above breakeven. In the report itself this is qualified by saying that "present day fusion experiments have already exceeded conditions equivalent to a $Q = 1$ operating power plant", without of course explaining what 'conditions equivalent' actually means. From the discussion in Chap. 9 we know that it means by extrapolating D-D results to D-T results, so-called extrapolated breakeven, which has nothing to do with reality. Moreover, a $Q = 1$ device cannot be called a power plant, as it is just a power consuming device. The statement made above may arguably be included in a publicity or advertisement leaflet but has no place in a scientific journal.

They make matters worse by stating in the body of the report that it "appears realistic that fusion power plants delivering electricity will be available for commercial use towards the middle of this century". The statement dates from 2005 when construction of ITER had not even started (the squabble about the ITER site had just ended) and the authors must have known that there was no scientific basis for making such a statement; they cannot themselves have believed it!

The rosy pictures painted by the leaders of the various fusion enterprises have in general been accepted unquestioningly by the media, as is often the case with science. Scientists still tend to be trusted for no other reason than that they are scientists. The media often even tries to surpass them in screaming headlines of great progress made towards unlimited energy or similar nonsense. There have been pitifully few who have not fallen for these fairy tales. One of these few and especially noteworthy is Steven Krivit, who runs the *New Energy Times* website. After having thoroughly debunked the cold fusion fiasco, from 2016 Krivit started to cast a critical look at the statements of leaders of hot fusion projects like JET and ITER. What he discovered was really shocking: a marvellous and almost endless collection of false statements, some so blatantly untrue that, assuming that the people who made them are competent in the field, they must have been made with the clear intent to lie and/or mislead the people they were addressing. I believe Krivit started with "The ITER Power Amplification Myth" in which he pointed out that the ITER management and communications office have led journalists and the public to believe that, when completed, the reactor will produce 10 times more power than goes into it. We know by now that that is not true, but that is only so because Steven Krivit did not tire of pointing this out. The true picture (from his website) is shown in Fig. 16.2. It shows that the amount of electric power that goes into the reactor will be 300 MW and that 536 MW may be produced in thermal power. If converted into electricity at 40% efficiency (which is high), it would result in 214 MW electric power output, leaving a shortage of 86 MW.

This stands in glaring contrast to what the ITER website stated in 2017: "The goal of the ITER fusion program is to produce a net gain of energy and set the stage for the demonstration fusion power plant to come. ITER has been designed to produce 500 MW of output power for 50 MW of input

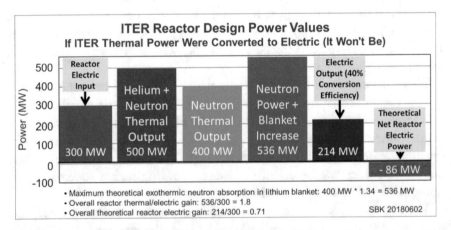

Fig. 16.2 Showing the electric power that goes into the ITER reactor, and the output in thermal power, resulting in net power usage, not power gain

power—or 10 times the amount of energy put in. The current record for released fusion power is 16 MW (held by the European JET facility located in Culham, UK)."

Thanks to Krivit's prodding, the text on the website has now been changed into: "ITER has been designed for high fusion power gain. For 50 MW of power injected into the Tokamak via the systems that heat the plasma it will produce 500 MW of fusion power for periods of 400–600 s. This tenfold return is expressed by $Q \geq 10$ (ratio of thermal output power to heating input power). The current record for fusion power gain in a tokamak is $Q = 0.67$ held by the European JET facility located in Culham, UK, which produced 16 MW of thermal fusion power for 24 MW of injected heating power in the 1990s."

In Chap. 9 we have already seen what JET's 16 MW of produced power actually means for the efficiency of the machine (just 2%, as also dug out first by Krivit, but kept under the rug by the JET people and only released when asked for, while it is the only number that really counts!). In Chap. 10 we also already related what ITER's 500 MW actually means and that the website of the Japanese JT-60SA has never thought it necessary to hide the truth.

The new text on the ITER website is still not a model of clarity. No member of the public will understand it without further explanation. It is nonetheless an improvement, although still untrue. ITER has not been designed for high fusion power gain, but for pitifully low power gain. After all, it is, as the JT-60SA website states, a zero (net) power reactor. It remains common policy among fusion managers not to mention how much electricity ITER requires to operate, and it is very difficult to figure this out. For instance, the 300 MW mentioned in Fig. 16.2 does not include ITER's non-interruptible power usage; power that will be needed regardless of whether the reactor is operating or not. We will come back to this below.

A wrong statement on a website is one thing but providing confusing or misleading information to the US Congress is quite another. However, as recently as March 2018 Congress was shamelessly misled by ITER director Bigot and the acting associate director of the Department of Energy's Office of Fusion Energy Sciences Van Dam, as Krivit reports. Such misleading statements tend to propagate through the community. Others start to repeat these inaccuracies on their own website. The wrong information spreads as a highly contagious virus and is almost impossible to eradicate, as a selection on Krivit's website shows. Even reputable journals like *New Scientist* do not seem capable of getting it right or providing correct information.

Krivit's comments and criticism are of course belittled with statements like "this discussion is irrelevant in the case of ITER since its purpose is not to produce as much energy as possible but to demonstrate the technological feasibility of fusion." If so, why do the fusion leaders mention these numbers and why don't they tell Congress to stop asking these irrelevant questions?

Krivit has much more to say about the probably deliberately deceptive way ITER is presented to the public. Further details can be found on the website of the *New Energy Times*. I will only repeat his conclusion here: "Given the preponderance of misrepresentations of the ITER power values on prominent Web sites, in news outlets such as the *New York Times*, Bloomberg, and the BBC, in science publications such as *Nature*, in major worldwide Web references such as *Wikipedia*, EUROfusion, and World Nuclear Association, and in a publication of the European Parliament, logical conclusions are that: (1) The fusion representatives who created the misrepresentations had to have known of the effects of their public relations efforts; (2) A significant number of fusion scientists who were not directly responsible for the creation of the misrepresentations must have read about their project in the news media and known of the falsehoods—yet for at least five years before October 2017, they corrected none of the falsehoods. Even the director-general of ITER, Bernard Bigot, had to have seen the falsehood in [an article in *Nature*], which says that ITER "is predicted to produce about 500 megawatts of electricity." He added a comment to the article after it [was] published. Was the public broadly misled? The list above shows that, yes, it was."

Lawrence Lidsky

The first criticism that aroused some attention goes back to a 1983 paper by the fusion insider Lawrence Lidsky (1935–2002), at the time a professor of nuclear engineering at MIT and associate director of the MIT Plasma Science and Fusion Center. Lidsky wrote a paper, called "The Trouble with Fusion" in the *MIT Technology Review*. He was of the opinion that the fusion programme had come prematurely under the sway of machine-builders and that as a consequence science was suffering. An explicit goal was established, generating commercial electricity from D-T fusion early in the twenty-first century. Once such a goal has been set, it is not easy to change. Producing net power from fusion is a valid scientific goal, but generating electricity commercially is an engineering problem, Lidsky said.

In the article he discussed all the problems that are still harassing nuclear fusion reactor designs based on the tokamak concept and trying to burn deuterium and tritium. He pointed out that the engineering problems involved in a fusion reactor are so immense that, even if a reactor is eventually built, no one will want to buy it, no power company would ever want to work with a reactor with such complex engineering. "The costly fusion reactor is in danger of joining the ranks of other technical "triumphs" such as the zeppelin, the supersonic transport, and the fission breeder reactor that turned out to be unwanted and unused." "A chain of undesirable effects ensures that any reactor employing d[euterium]-t[ritium] fusion will be a large, complex, expensive and unreliable source of power." He pointed out that deuterium-tritium as a fuel had undesirable effects, including a tritium supply problem and the release of high-energy neutrons, whose bombardment would weaken the reactor structure and make it radioactive, problems that still need solving now in the early twenty-first century when we should have been enjoying commercially attractive electric power from fusion reactors. He also pointed out that nuclear fusion reactors would inevitably be very complex as they had to handle enormous heat flows and huge temperature gradients over a distance of a few metres (from the 150 million degrees of the plasma to a few degrees above absolute zero for the superconducting magnets). When he formulated his criticism, tokamaks were still fairly small, with a couple of large ones (TFTR and JET) in the process of being built, and they are becoming ever larger, with costs going through the roof.

At the time he was not thanked for his message. On the contrary the reaction was swift and substantial, leading to Lidsky being quietly 'purged' in almost Soviet style, made a pariah in the fusion community, and stripped of his title as an associate director of the MIT Plasma Fusion Center.[1] In particular, Harold Fürth, the then director of PPPL, went to great lengths to try to destroy every argument Lidsky was making and trying to nip in the bud any damage his article might have caused. TFTR was being constructed, but its successor was being planned. It would be the first device to achieve ignition, but was never built, as Congress slashed the fusion budget. The decrease was not that much, just five per cent, but devastating for these plans for which an increase was needed. It may well be that Lidsky's dissent affected the vote in Congress.

Reading Lidsky's paper now, more than thirty-five years after it was published, it is clear that his arguments are still as fresh and apt as they were at the time, even more so as solutions are more urgently needed. A fusion reactor will indeed be large, complex and expensive. The neutrons are a problem and tritium breeding is not that simple. Since no reactor has yet produced any

---

[1]Lidsky was asked to resign (a questionable request for an academic institution to make) by Ronald C. Davidson (1941–2016), the first director of the MIT PSFC, who later became director of PPPL.

power to speak of, it is not yet possible to say whether it will be unreliable, but the signs are certainly not favourable.

After his rather early death in 2002 at the age of 66, one of Lidsky's colleagues observed: "He was one of the earliest engineers to point out some of the very, very difficult engineering challenges facing the program and how these challenges would affect the ultimate desirability of fusion energy. As one might imagine, his messages were not always warmly received initially, but they have nevertheless stood the test of time." They still do.

Robert L. Hirsch

Hirsch, whom we encountered in Chaps. 8 and 9, was director of the US fusion programme in the 1970s, when he aggressively pursued the tokamak option. He even predicted that commercial fusion power could be on the grid by the year 2000. However, he did not take long to drastically change his mind. Already in 1985, he infuriated the American fusion community by denouncing the tokamak as an impractical reactor design. The fatal flaw of the tokamak, he said, is that "it is inherently a complex maze of rings and a toroidal chamber inside of other rings. In my view, this complex geometry will not be acceptable to the utility world, where power plants must be maintained and serviced rapidly at low cost. In that world, simple geometries are essential." With this he was essentially making the same point of unacceptable complexity that Lidsky had made earlier.

Twenty-five years later in a book on the history of the US fusion effort, Hirsch elaborates on his earlier remarks by stating that the current effort to achieve commercial fusion energy will almost certainly fail for three main reasons: (1) fusion power plants as currently envisaged will be extremely large and complicated, and consequently very expensive while low cost is a major requirement of electric power generation; (2) the radioactivity problem will require complicated remote handling operations, expensive maintenance and disposal of large volumes of radioactive waste (see Chap. 18); and (3) the currently envisaged fusion reactors are not inherently safe (see also Chap. 18). They require large high-field superconducting magnets, which contain large amounts of stored energy. There is a small, but nevertheless possible risk for this energy to be explosively released, if one or several of the magnets were suddenly to become non-superconducting.

And Hirsch has stuck to this view as a few years back he wrote in *Science and Technology*: "tokamak fusion power will almost certainly be a commercial failure, which is a tragedy in light of the time, funds, and effort so far expended." He even wants to learn lessons first from the tokamak experience before embarking on another method, quite a different attitude from the one he himself displayed when directing the AEC fusion programme. His cavalier approach, which seems pervasive in US fusion research, has damaged the fusion enterprise more than anything else, I believe. Everybody knows that science cannot achieve everything one would like it to, and certainly not on command or by just throwing a lot of money at it. Certain things are just not or not yet possible, because of physics or engineering problems or simply for lack of knowledge and understanding of the physics behind them. It makes sense to try to find out first what the basic problems of a certain approach are before embarking on large-scale and hideously expensive ventures. Of course, if Hirsch's prediction of fusion power on the grid by the year 2000 had come true, he would have been hailed as a visionary and the saviour of mankind, so for him perhaps the risk was worthwhile although he must have known that the odds were heavily against him.

Criticism of ITER

Let us start this section with Hirsch who of course had also something to say about ITER. In a book he wrote with two others in 2010, entitled "*The Impending World Energy Mess*", in which as the title suggests impending doom is prophesized (mainly as regards an oil shortage), it is said "the outlook for success with the physics aspects of ITER is good, but the likelihood of commercial success is near zero. Thus, when ITER operates 'the operation might well be a success, but the patient will in effect be dead'. As a result, the world will have wasted decades and tens of billions of dollars on a dead-end concept. Sadly, the ITER waste could have been avoided." For the latter remark Hirsch is advised to look in the mirror and see the man who is (partly) to blame for this.

Hirsch, however, gets too much credit as a critic. There are much more capable people, very capable people indeed, on a par with Lidsky, who have joined the critics: the French physics Nobel Prize winners Pierre-Gilles de Gennes (1932–2007) and George Charpak (1924–2010) and the Japanese physics Nobel Prize winner Masatoshi Koshiba (1926–2020), who have in particular criticized ITER as a useless and overpriced reactor. De Gennes spent part of his career with the French Atomic Energy Commission (CEA) and can be considered an expert in all things nuclear. He felt that too much money is spent on actions that are not worth it and mentioned nuclear fusion as an example: "European governments, as well as Brussels, have rushed into the ITER experimental reactor without having carried out any serious reflection on the possible impact of this gigantic project. (...) We are unable to fully explain the instability of plasmas or the thermal leaks of current systems. So, we are embarking on something which, from the point of view of a chemical engineer, is heresy. (...) The ITER project was supported by Brussels for reasons of political image, and I find that it is a fault."

Georges Charpak was a member of the CERN staff since 1959 where he invented and developed the multiwire proportional chamber. Charpak's main criticism concerns the enormous costs of ITER, in combination with the fact that in all likelihood it will not solve the problems fusion is faced with ("it will only study the stability of its own plasma"). He had earlier vented his concern that ITER's cost would eat into the science budgets of EU member countries. In a last interview in 2010 just before his death and just before construction of ITER started in earnest, he and his colleagues Jacques Treiner and Sébastien Balibar said that what they had feared was actually happening: "the estimated cost of the construction of ITER has gone up from 5 to 15 billion euros (it is even more now (LJR)), and there is talk of passing on the consequences to the European budget for scientific research." "This is exactly the catastrophe we feared. It is high time to give it up. (...) Much more important research, including for the energy future of our planet, is thus threatened." They continued by stating the three major problems for fusion that nobody has managed to solve for more than 50 years: to keep the plasma inside the reactor vessel, as it is unstable; to produce tritium in industrial quantities and to invent materials to enclose this plasma in a vacuum vessel of a few thousand cubic metres. The most formidable problem, they said, is the third: violently irradiated by very energetic neutrons (14 MeV) emitted by fusion reactions in the plasma, the material of the vessel will lose its mechanical strength. No matter how often it is said that materials can be imagined that will resist irradiation, they remain sceptical. The development of a prototype power plant, let alone a commercial power station, is still very far away. It is in no way justified to take money away from other research projects on the pretext that an almost infinite source of energy is on the horizon. Plasma physics must be funded on a par with the other major fields of fundamental research. They then say: "If we continue, all areas of research will suffer. This situation is reminiscent of the construction of the International Space Station (ISS). Another pharaonic project, the ISS cost $100 billion and our colleague astrophysicists still remember the budget cuts that its construction entailed. What was the ISS used for? For virtually nothing. To observe the Earth or the Universe, it is better to send robots into orbit that are more stable and cheaper. In fact, the astronauts are bored up there. So, they spend their time studying their own health! ITER is likely to be comparable: if built, this large machine will only be used to study the stability of ITER's plasma. Isn't it a little expensive to spend 15 billion euro on that? (…) So, rather than masking an initial bad decision by an even worse escalation, it would be better to finally admit that the gigantism of the project is disproportionate to the expectations, that its management appears deficient, that our budgets do not allow us to pursue it, and transfer that money to useful research."

There is no way around their arguments. Particularly unconvincing is the claim that, since ITER is now funded directly from the EU budget, French and European researchers would be immune from budgetary restrictions caused by ITER. A very weak argument indeed as money can be spent only once and the money spent on ITER cannot be spent on other (energy) research or climate-change combating measures, of which there are currently pitifully few. Many EU member countries carry out ITER-related work, which is, in any case partly, funded by these member countries, money that, if there were no ITER, would be spent on other scientific projects. There may be a division between ITER funding and funding for other research at EU level, but such division does not exist at EU member state level, so spending for ITER undoubtedly affects other scientific research in EU member states. The same is true for the other ITER Members. For the US, for instance, ITER is eating the greater part of the budget of the Office of Fusion Energy Sciences. Because of ITER's thirst for cash, many other concepts have been strangled or shut down. Researchers often need to "show relevance" to ITER, otherwise they run the risk of not obtaining funding or of getting closed down.

Koshiba, too, issued very apt warnings about the highly energetic neutrons that will be released in huge quantities in the fusion reactions and that scientists do not know how to handle. How to absorb them by the walls surrounding the reactor, without the material becoming radioactive and forcing a biannual replacement for which the reactor will have to be shut down, an expensive and uneconomical solution? No wonder Koshiba was very happy when it was decided to build ITER in France and not in Japan.

A remarkable example of a retired fusion scientist who is beginning to speak out against the ITER folly is Daniel Jassby, a principal research physicist at the Princeton Plasma Physics Laboratory until 1999. He stood at the cradle of the spherical tokamak and is clearly not someone who can or should be dismissed out of hand, not a 'mere' journalist. In a few publications in the online version of the Bulletin of the Atomic Scientists and on the website of the American Physical Society he made short shrift of ITER and the entire fusion enterprise. He is worth quoting in some detail.

Above we have already discussed the power balance at ITER, but that only concerned the actual fusion experiment. Not included in that calculation is ITER's non-interruptible electric power drain which varies between 75 and 110 MWe, i.e., power that has to be drawn from the main grid continuously, day and night, even when the plant is not operating. Even during the coming years of plant construction, the on-site power consumption will average at least 30 MWe. When ITER actually operates, its plasma will require at least 300 MWe, the output of a small power station, for tens of seconds for heating and generating the necessary plasma currents. During the 400 s operating phase, about 200 MWe will be needed to maintain the fusion burn and control the plasma's stability. The 50 MW of heating power injected into the plasma helps sustain its temperature and current, and is only a small fraction of the overall electric power input to the reactor, which varies between 300 and 400 MWe. One wonders what the great difficulty is for the ITER Organization to give an honest picture of the energy balance on its website, instead of (falsely) boasting of being "the first of all fusion experiments in history to produce net energy gain."

A fundamental question is whether ITER will produce 500 MW of anything, a query that revolves around the vital tritium fuel—its supply, the willingness to use it, and the campaign needed to optimize its performance. Fusion practitioners, so Jassby states, are in fact intensely afraid of using tritium for the two reasons we have already seen: its radioactivity, which implies that there are safety concerns connected with its potential release into the environment, and the unavoidable production of radioactive materials as fusion neutrons bombard the reactor vessel, requiring enhanced shielding that greatly impedes access for maintenance and introduces radioactive waste disposal issues. Assuming that the ITER project is able to acquire an adequate supply of tritium and is brave enough to use it, nobody knows whether it will actually achieve 500 MW of fusion power. ITER's current schedule envisages the use of deuterium and tritium in 2035. But there is no guarantee of hitting the 500 MW target. During the unavoidable teething stages through the early 2040s, it is likely that ITER's fusion power will be only a fraction of 500 MW, and that more injected tritium will be lost by non-recovery than actually burned. The permeation of tritium at high temperature in many materials is not understood to this day. The deeper migration of some small fraction of the trapped tritium into the walls and then into liquid and gaseous coolant channels cannot be prevented. Most of such tritium will eventually decay, but there will be inevitable releases into the environment via the circulating cooling water. In designs of future tokamak reactors, it is commonly assumed that all the burned tritium will be replenished by tritium breeding in the lithium blanket surrounding the plasma, i.e., that they will be self-sufficient in tritium. But that fantasy, too, totally ignores the tritium that is permanently lost in its globetrotting through reactor subsystems. It may well be that in ITER the total of unrecovered tritium rivals the amount burned and can be replenished only by the costly purchase of tritium produced in fission reactors. The conclusion must be that tritium self-sufficiency is a fantasy, not only for ITER, for which self-sufficiency in tritium is not a requirement, but also for DEMO or real fusion power plants. We stated before that there will be no commercial development of fusion energy if self-sufficiency in tritium cannot be achieved, so the conclusion from this is clear.

Whether ITER performs well or poorly, according to Jassby, its most favourable legacy is that, like the ISS, it will have set an impressive example of decades-long international cooperation among nations that are both friendly and semi-hostile. But why would cooperation for the sake of cooperation be a good thing? And if one wants at any price to stimulate international cooperation, then better to choose something that is bound to be useful, such as the eradication of some ghastly disease. Judging from the attitude of various countries during the current Covid-19 pandemic, the ITER cooperation effort has not left behind any obvious desire for more collaboration and understanding.

Moreover, the ITER cooperation would not have got off the ground if the EU had not foolishly agreed to pay about half the cost, letting the other participants enjoy a cheap ride. The international collaboration and the resulting immense management problems have greatly amplified ITER's cost and timescale. All nuclear energy facilities—whether fission or fusion—are extraordinarily complex and exorbitantly expensive. Other large nuclear enterprises have experienced a tripling of costs and construction timescales that ballooned from years to decades. And yet ITER will beat the lot of them with its estimated price of $60 billion. The only sensible conclusion from this can be that all such projects should be abandoned as soon as possible. They cannot be the solution! If anything, they are part of the problem.

ITER may allow physicists to study long-lived, high-temperature burning D-T plasmas. As such, ITER will be a havoc-wreaking neutron source fuelled by tritium produced in fission reactors and powered by hundreds of megawatts of electricity from the regional electricity grid, demanding unprecedented cooling water resources.

Structural damage in ITER will not exceed 2 dpa at the end of its rather short operational life, but that will be quite different in any subsequent fusion reactor that attempts to generate enough electricity to exceed all the energy sunk into it. That reality alone should be enough to abandon fusion as an energy generating option. As Jassby states, rather than heralding the dawn of a new energy era, it is instead likely that ITER will perform a role analogous to that of the fission fast breeder reactor, whose blatant drawbacks mortally wounded another professed source of "limitless energy" and enabled the continued dominance of light-water reactors in the nuclear arena.

Doubts are also being voiced by people still working in the fusion enterprise, albeit still rather timidly and mainly by people who have another interest. In this respect Leonid Zakharov, a leading plasma physicist at PPPL and currently a proponent of some type of spherical tokamak, is a fairly vocal critic who stated at the Physics Colloquium at Princeton University as far back as December 2000 that "tokamak fusion devices (…) are now in an eventual state of defeat and possible shutdown in the US. Despite much better understanding of the tokamak plasma now, many fundamental problems on the way to the tokamak reactor remain unresolved even at the conceptual level. These problems include stability and steady-state plasma regime control, power extraction from both the plasma and the neutron zone, activation and structural integrity of the machine under 14 MeV fusion neutron bombardment, maintenance of future reactors, etc." This he wrote in 2000 and, as nothing much has changed since then in respect of these problems, he has consistently propagated the same message. In a talk in 2018, in answer to the self-posed question "Can expectations (i.e., the expectations for fusion to happen) be converted to reality?" he states that "from the mid 1980s to 1990s insufficient attention to science was shown for addressing fusion reactor problems. Then TFTR and JET failed. The leaders disappeared, the program became complicated and unmanageable, and progress was lost. Nobody was capable of understanding that the failure of TFTR and JET indicated that the adopted approach was exhausted. ITER is the implementation of the same failed approach. There are many indications that the program is in the stage of degradation being insensitive to science and experimental data."

Finally, in a very recent paper he writes: "In 2020 (now 2021 (LJR)) the world best tokamak JET will perform the second D-T experiment in the same high recycling regime, which already failed in 1997. With no way of getting $Q = 1$ or even 0.6 as in 1997, it will provide the experimental evidence that the currently adopted approach to fusion is incapable of making progress and is hopeless. It was exhausted 20 years ago. The JET D-T experiment will be proof of the failure of the entire fusion crowd, including management, who were ignorant of science and relied instead on interpretations, scaling, cooked up explanations, and fake understandings. Since the mid-1990s, the science leaders of the program have disappeared and science itself has become unwelcome in the program."

All this is pretty devastating. His statement confirms the assertion that the whole enterprise went down hill when the fusion project was 'upgraded' from a scientific research project to an engineering project that tried to take a shortcut to a power generating facility, a shortcut that has now landed it in a cul-de-sac.

Criticism of Other Fusion Endeavours

Magnetic fusion research today focuses almost exclusively on the tokamak concept. As we have seen, alternative concepts are at present mostly pursued by privately funded companies. Jassby has branded many of them as pursuing Voodoo Science. This expression for scientific research that falls short of adhering to proper scientific methods was coined by the American physicist Robert L. Park (1931–2020) in his book *Voodoo Science: The Road from Foolishness to Fraud*. He names cold fusion and the ISS as examples of such science. So, how can such an exalted exercise as trying to "bottle the Sun" be Voodoo Science?

As we have seen, many of these start-ups promise to develop practical electric power generation from fusion within 5–15 years, and claim to do so by surpassing ITER's planned performance in a fraction of the time and at 1% of the cost. According to Jassby, these projects are nothing more than modern-day versions of Ronald Richter's scam of 1951, the first of its kind in fusion history. In that year Argentinian president Perón announced to the international press that "the Argentine scientist Richter"—a German who couldn't speak a word of Spanish—had achieved the controlled release of nuclear-fusion energy. It soon became evident that the claims were spurious, but it jolted other countries into starting fusion projects. Just as Richter's contraption was unable to generate a single fusion reaction, none of the current projects has given evidence of more than token fusion-neutron production, if any at all.

Park demolished "cold fusion" but never mentioned any of the failed "warm plasma" fusion schemes of his era in either his book or any of his columns. Most of these plasma-based fusion attempts can nonetheless be classified as voodoo technology, voodoo fusion, defined for present purposes as those plasma systems that have never produced any fusion neutrons, but whose promoters claim will put net electric power on the grid in just a couple of years from today. The messianic incantations of the voodoo priest-promoters, so Jassby says, invoke the aura of "the energy source that powers the sun and stars" as well as the myth that terrestrial fusion energy is "clean and green" in order to cast a spell over credulous investors and politicians. The point is that more than 90% of fusion concepts have never produced measurable levels of fusion neutrons, implying that those systems may have little practical value.

Jassby excludes some efforts, like those of Tokamak Energy and Commonwealth Fusion Systems from the voodoo class, despite what he calls their preposterous and unjustifiable claims of near-term electric power production, solely because their schemes are based on tokamaks. Tokamaks have demonstrated for 50 years that they can produce fusion reactions, which is of course quite another thing from producing net energy that can be turned into electricity to feed the grid.

Conclusion

All this criticism is mostly being ignored, presented as false claims or miscommunication, or belittled as nothing new. The juggernaut just thunders on, crushing everything and everybody on its path. In that respect it is very convenient that much of the criticism comes from old(er) people after their retirement from active research. Fusion is a long-term project and these critics conveniently die and can then be safely ignored. I bet that hardly anybody now working in the fusion community will for instance have heard of Lidsky's criticism, even though it is still very much to the point and was hitting the nail on the head close to forty years ago. Even more than that: he has been completely vindicated. When all finally come to their senses and realise the folly of it all, it will hit science hard. It will cause permanent damage to the entire science enterprise and bring home to society that scientists are as unreliable as the rest of us, and just put their mouth where the money is. It is abundantly clear that the fusion enterprise as currently conducted is a hopeless one.

Let me finish this chapter with a quotation from a 2018 interview with Chris Llewellyn Smith, former director-general of CERN and of the JET facility at Culham (UK), and currently Director of Energy Research at Oxford University. He has always been a staunch proponent of nuclear fusion, but now seems to be slowly changing his mind and is becoming more critical: "Ten years ago, I would have replied that I'm reasonably confident that we will be able to make a fusion power plant, although we need ITER to be sure, and the real question is can we make one that's reliable and competitive? The question of reliability will remain unanswered until we try, although operation of ITER will provide clues. I used to think there was a reasonably good chance that fusion could compete with other low carbon sources of power, but, while I would not say that it is impossible, the situation has changed. The cost of wind and solar power has decreased faster than anyone could have dreamt. Meanwhile ITER has gone way over budget, partly because of the way that the project was set up and because it's the first of a kind, but probably also because fusion reactors will be intrinsically more expensive than we thought a decade ago. I think we need to finish ITER and establish once and for all whether fusion really is a viable option. We will then have to reassess the likely cost of fusion power in the light of the experience gained with ITER and in comparison with the cost of alternatives before deciding whether to go ahead and build a real fusion power station." We are now two years later, and it seems to be a foregone conclusion what the result of the reassessment referred to here will be.

17

Economics and Sustainability

All the expense and effort on fusion experiments will only be justified when in the end a viable reactor emerges that produces electricity reliably and at a competitive rate. In this respect, various studies in the last thirty years have already indulged in investigating the potential for fusion power, but always by people for whom a lot was at stake; insiders in the field whose jobs and research opportunities depended on a favourable outcome for the prospects of fusion. In a 2005 European study on commercial fusion power plants, for instance, it was concluded based on very little hard data that "fusion power has *very promising* potential to provide inherent safety and favourable environmental features, to address global climate change and gain public acceptance. In particular, fusion energy has the potential of becoming a clean, zero-CO_2 emission and inexhaustible energy source." Knowing that that on its own would not be enough it was added that "the cost of fusion electricity is likely to be comparable with that from other environmentally responsible sources of electricity generation." Especially objectionable in this conclusion are the words 'very promising', as these were not based on any facts. JET and TFTR had just had some very limited success with the first D-T operations (less than expected as both were supposed to achieve breakeven) and the rest was just wishful thinking. The second conclusion is in any case completely false, as the cost of nuclear fusion power plants will in all probability be exorbitant and the costs of 'environmentally responsible sources' like solar power and wind have substantially decreased. In this connection it is not helpful that the world is awash with oil and gas, and that in the wasteland of the

L. J. Reinders, *Sun in a Bottle?... Pie in the Sky!*, https://doi.org/10.1007/978-3-030-74734-3_17

free market this will be the case for a considerable time to come, resulting in dirt-cheap electricity from fossil fuels. Although it is clear that these resources are finite and will not last forever, predictions of peak oil, peak gas and peak coal to be reached in the first decade of this century have all turned out to be false by a wide margin.

Renewable sources, wind and solar, are on the march, as climate change has brought home the urgency of switching from fossil fuels to carbon-free energy sources. A carbon tax is undoubtedly one of the most efficient ways to achieve this. Some sort of price on carbon has now been adopted by more than 40 governments worldwide, either through direct taxes on fossil fuels or through cap-and-trade programmes. The European Union Emissions Trading System (EU ETS) is such a cap-and-trade system. It is the biggest greenhouse gas emissions trading scheme in the world and sets a maximum on the total amount of greenhouse gases installations may emit. "Allowances" for emissions are then auctioned off or allocated for free, and can subsequently be traded. The system has been criticized for several failings and in general for failing to meet its goals. It cannot be considered as a proper means to deal with the problem. In practice, most countries have found it politically difficult to set prices that are high enough to spur truly deep reductions in fossil fuel use. Carbon pricing programmes are mostly fairly modest and partly for that reason, carbon pricing has, so far, hardly played any role in efforts to mitigate global warming.

The International Monetary Fund has recently concluded that a carbon tax should be imposed immediately around the world and should rise to $75 per tonne of CO_2 by 2030. If imposed, it would add about $0.17 to the price of a litre of gasoline.[1] The average carbon emission allowance price in the EU ETS, which is determined by the market, increased from €5.8 per tonne in 2017 to around €25 per tonne in 2020, still well below the price stated by the IMF report. In the end such tax is the only way to have users of carbon fuels pay for the climate damage they cause by releasing carbon dioxide into the atmosphere. It will ensure a level-playing field by factoring in all the costs in a proper way, and it can then be left to the market to make the technology choices.

Even if such tax were levied, it would make little difference for fusion as it will not play a role of any significance in energy generation, at any rate not in this century. One of the first economic impediments in making energy from fusion economically attractive is the unprecedentedly high level of investment needed for the proof of principle (the cost of ITER and DEMO) and the

---

[1]One litre of gasoline weighs 0.75 kilo and produces 2.3 kg of CO_2. The proposed tax on 2.3 kg would be 2.3 times $0.075 which equals $0.1725 per litre of gasoline.

long construction time of fusion plants. Within the mainstream scenario of a few DEMO reactors towards 2060 and the subsequent construction of a few relatively large reactors, there is no realistic path for fusion to make an appreciable contribution to the energy mix in this century.

In other words, fusion will not contribute to the energy transition in the time frame of the 2015 Paris climate agreement, i.e., to achieve a near complete decarbonization of energy generation by 2050, and even if this is delayed for a couple of decades the situation for fusion will not markedly improve.

But leaving these details aside we can still attempt to calculate or reason about whether a fusion power plant (after being built) can indeed be competitive with, for instance, wind and solar energy. The latter two energy-generating options are also still in development and several problems have to be solved before they can make a real impact. But, they are actually already generating huge quantities of energy, which is the reason why the investment in solar (from solar panels) and wind power soared to $288.9 billion in 2018, approximately 3% of the world energy market, with the amount spent on new capacity far exceeding the financial backing for new fossil fuel power. It was the fifth year in a row that investment exceeded the $230 billion mark. In the light of such figures, fusion with a global investment of a mere $2–3 billion per year is just a pitiful exercise. It will never be able to match such numbers as long as it does not produce any net energy. But it also implies that fusion will not be the long-awaited saviour that rescues the world from an impending carbon-dioxide death. Whenever it becomes available, it will enter a largely decarbonised market that will have organised itself in one way or another. An intruder or newcomer in such market must offer something that is not already available, or offer it more cheaply, or be competitive in another way. Being carbon free, safe, clean and unlimited will not be enough. This is nowadays demanded from any new energy source. Wind and solar are also carbon free and safe, and even more unlimited (no tritium supply problem for instance) and cleaner than fusion (no radioactivity whatsoever).

The most basic problem of solar and wind power is that they do not always generate energy at the right time. Reliable ways to store the generated energy must be found, which is by no means a simple matter. It is at present still necessary to complement basic grid needs with fossil fuel power. Such problems are, however, peanuts compared to the problems fusion is faced with, where a proof of principle is still far away.

Two solutions for the storage problem are currently being explored and each will probably provide part of the solution. The first is to develop more efficient batteries, and the second solution is to use the generated energy to produce hydrogen through the electrolysis of water. The hydrogen is then stored and can later be used, burnt in a power station for instance, for the reverse process whereby energy is released.

Is Fusion Energy Sustainable?

According to the stories told on the websites of fusion promoting endeavours, fusion power seems like the perfect energy source. They say that it is clean, inexpensive, and can draw from an inexhaustible resource of fuel. But is this actually true? In the preceding chapters we have already made some critical remarks about this story. It is true that deuterium is almost free, but that is certainly not the case for tritium, and both are needed in the envisaged nuclear fusion power stations. The prospects for procuring enough tritium, or for breeding enough tritium in the reactor itself, look very bleak.

The biggest sources of tritium are Canada Deuterium Uranium (CANDU) nuclear fission reactors that use heavy water (D_2O) as moderator (to reduce the speed of the fast neutrons produced in the fission reactions). Today there are 31 CANDU reactors in use around the world. Apart from Canada, which is the world's largest producer of tritium with 19 such reactors, they are operational in Argentina, India, Pakistan, China, South Korea and Romania. A typical CANDU reactor produces about 130 g of tritium per year.

The tritium required for ITER (12.3 kg) will be supplied from the production of the CANDU nuclear reactor in Ontario. The tritium required to start DEMO will depend on advances in plasma fuelling efficiency, burnup fraction, and tritium-processing technology. In theory it is possible to start up a fusion reactor with little or no tritium, but at an estimated cost of $2 billion per kilogram of tritium saved, it is not economically sound. If ITER and further fusion developments are successful, two or three countries may build their own fusion reactors, all requiring tritium from these same sources. If Canada, Korea and Romania make their tritium inventories available for fusion, there is a reasonable chance that 10 kg of tritium would be available for fusion R&D in 2055, but stocks would likely have to be shared if more than one fusion reactor is built. There simply will not be enough tritium.

This reflects the point already made several times in this book that tritium breeding is of paramount importance for fusion ever to become a success. For the tritium-breeding blankets to be used in future reactors, lithium and some other metals are needed, and their availability is not without problems either.

Lithium, Beryllium and Lead

Let us first list some facts about lithium, which as regards fusion is the most important of these elements. The chemical element lithium (Li), with atomic number 3, is the lightest of all metals. It is a so-called alkali metal, i.e., belonging to the same group as potassium (K) and sodium (Na), and widely distributed on Earth. Due to its high reactivity it does not naturally occur in elemental form and is always found bound in stable minerals or salts. It appears in the form of two natural isotopes: ^{6}Li (7.42%) and ^{7}Li (92.58%). Lithium has many uses, the most prominent ones being in batteries for cell phones, laptops, and electric and hybrid vehicles (46% in 2019).

In the Earth's crust (upper 16 km), lithium can be found in the range from 20 to 70 ppm by weight in various forms. In the first place, minerals with concentrations of Li_2O (lithium oxide) from about 4.5 to 7%. The most important mines are in Australia. The second form is brine, which is a high-concentration solution of salt in water. The evaporation process of salt lakes leads to an increased lithium content and, when the lake is completely dried up, the remaining salt can contain 4–6% of its weight in lithium. The lithium is extracted as lithium carbonate (Li_2CO_3). The present most important production site is in Chile. The world's biggest salt flat is located in Bolivia. It holds perhaps 17 percent of the planet's total lithium reserves and will soon be put into production. The third and most important potential source of lithium is seawater, which has a mean lithium content of 0.17 ppm. The world's oceans contain an estimated 180 billion tonnes of lithium. Up to now, only experimental processes have been tested for extracting lithium from seawater, but the production price is expected to remain too high to be competitive in the present market.

Fusion reactors will use lithium in the tritium-breeding blankets surrounding the plasma.

However, it is not so easy to estimate the required amount of lithium and the required enrichment level, i.e., its ^{6}Li content. To be able to make such an estimate, a relatively detailed knowledge of the blanket and—in the case of liquid breeders—the design of the infrastructure system, including all pipework, cooling, cleaning, and tritium extraction systems would be required. So far only concepts of blankets exist, some of which will be tested in ITER. It has been estimated that the lithium-lead blanket of a 2 GW (fusion power) device will require 8200 tonnes of a lead-lithium alloy (Pb–Li), enriched to 90% of ^{6}Li. This implies that 52 tonnes of pure ^{6}Li will be needed (26 tonnes per GW). The consumption of ^{6}Li depends on the tritium production rate (2 g of ^{6}Li is needed to produce 1 g of tritium) and is small compared to the large total lithium inventory (112 kg of ^{6}Li consumption per full power year and GW). For solid breeders, 'fusion grade' lithium (i.e., lithium with an isotopic composition that can be used directly in the blankets) refers to an enrichment level of 30–60%, for liquid breeders up to 90%. Such large quantities of enriched ^{6}Li will require an isotope separation process with a minimum output of several tonnes of 'fusion grade' lithium per full power year. There is currently no facility available that could satisfy this demand, and nor would it be straightforward to build such a plant. If DEMO becomes operational in the 2050s, blanket manufacturing must start in the mid-2040s. Several tonnes of enriched lithium will then already be needed, and a fully operational isotope separation plant must be ready by the late 2030s.

The ITER website paints a very optimistic picture of the availability of lithium. It asserts that lithium from proven, easily extractable land-based resources would provide a stock sufficient to operate fusion power plants for more than 1000 years. What is more, it says, lithium can be extracted from ocean water, where reserves are practically unlimited. Not a word about the costs or about other applications demanding a big share of the lithium pie. Another place on the website even states that "lithium availability will not be an issue for let's say the next thousand years, as there are approximately 50 million tonnes of proven lithium reserves in the world, which means about 3 million tonnes of ^{6}Li (7.42%) (…). It takes 140 kilos of ^{6}Li to obtain the 70 kilos of tritium necessary to produce one gigawatt of thermal power for one year. Assuming an availability of 80% and a conversion efficiency from thermal to electric power of 30%, the production of one gigawatt of electric power (the estimated size of an average fusion reactor) will require approximately 500 kilos of ^{6}Li per year, which would bring the total requirement for 10,000 reactors to 5,000 tonnes annually. Obtaining 5,000 tonnes of the precious isotope will require processing (…) approximately 70,000 tonnes of "regular" lithium… still a very small fraction of available resources. Fusion specialists generally consider that, in a world where all energy would be produced by fusion, the quantity of lithium ore present in landmass would be sufficient to provide the required tritium for several thousand years. As for lithium present in oceans, it could last us for millions of years."

The fact that there is no isotope enrichment plant available that can provide such amounts of ^{6}Li is not even mentioned, but also in other respects it is doubtful that the situation is as simple as depicted here. In the picture painted above, resources and reserves seem to be taken to mean the same, which is not the case, and several things have just been swept under the rug, notably the fact that it will not be possible to achieve self-sufficiency in tritium with ^{6}Li alone. As explained in Chap. 11, a neutron multiplier is needed, as ^{6}Li only produces tritium and helium when struck by a neutron. Using ^{7}Li would yield an extra neutron which can again react with another lithium nucleus to produce tritium, but that reaction requires energy. Without such extra neutrons self-sufficiency in tritium cannot be achieved. Therefore, beryllium is proposed as a neutron multiplier, causing further supply problems, as we will see below.

In any case, the 50 million tonnes mentioned for the reserves is questionable and 70,000 tonnes is not a small fraction of actually annually mined lithium, but close to the total amount mined in 2019 (77,000 tonnes of pure lithium). Estimates of lithium reserves (not counting the lithium in ocean water) differ vastly, from around 4 million tonnes to roughly 40 million tonnes.

However, when making such estimates a distinction must be made between lithium reserves and lithium resources, where the reserve estimates refer to the *extractable* portion of the resources. Estimates for lithium resources do indeed go up to more than 50 million tonnes, even as high as 80 million tonnes, but that would be a wrong number to quote for the reserves. In any case, it is surprising how large the differences between the various studies of lithium reserves or resources are. One reason for this is that most lithium classification schemes are developed for solid ore deposits, whereas brine is a fluid for which these schemes are less suitable, due to varying concentrations and pumping effects. Some of the differences also arise from the use of different numbers for the percentage of lithium contained in ore or salt. The amount estimated in 2018 by the US Geological Survey (USGS) is 14 million tonnes of economically recoverable lithium reserves worldwide (from over 39 million tonnes of lithium resources), while statistica.com reported a total of 15 million tonnes in lithium mineral reserves for the top nine countries in 2019.

It seems that in terms of lithium ore or salt deposits there is indeed no shortage on earth. Western Australia alone hosts five of the world's biggest lithium mines, whose combined resources exceed 475 million tonnes of ore, containing 1 to 3 per cent of lithium carbonate equivalent (LCE)[2], good for a couple of million tonnes of lithium. Knowing deposits of ore or salt and their lithium content, it is indeed easy to calculate reserves, but that does not yet mean that they can be mined in an economically and environmentally sound

---

[2]Data relating to lithium grades in mineral and ore resources and reserves for hard rock and brine deposits are reported using a number of differing measurement units, e.g., ppm Li and percentages of Li, Li_2O, or lithium carbonate. To normalise this data, it is often reported in "lithium carbonate equivalent" or "LCE", so that information can be easily compared on a like-for-like basis.

manner. The higher reserve estimates mentioned above also include reserves that are of no actual or potential commercial value.

No wonder that in the literature doubts have been raised about the adequacy of easily minable lithium deposits. The recent predictions of the future demand for lithium-ion batteries are alarming. For rechargeable batteries manufacturers use more than 160,000 tonnes of lithium every year, a number that is expected to grow nearly 10-fold over the next decade. It is conceivable that the automobile industry may acquire, and according to some estimates even use up, the terrestrial lithium reserves in the next few decades. Various battery configurations use between about 10 and 22 kg of lithium per car. Just imagine what that implies as regards lithium demand if a large proportion of the close to 100 million cars (2008 figures) manufactured annually are to be electric vehicles. Millions of tonnes annually just for car batteries. Nothing may be left for anything else.

A 2007 report, with a follow-up in 2008, entitled "*The Trouble with Lithium*" states "that there is insufficient *economically recoverable* lithium available in the Earth's crust to sustain electric vehicle manufacture in the volumes required, based solely on lithium-ion batteries. Depletion rates would exceed current oil depletion rates and switch dependency from one diminishing resource to another." Moreover, mass production of lithium carbonate is not environmentally sound, it will cause irreparable ecological damage to ecosystems making lithium-ion propulsion incompatible with the notion of the 'Green Car'.

Global demand for lithium is projected to go up from 307,000 tonnes in 2019 to 820,000 tonnes of LCE in 2025, while S&P Global estimates that new mines and brine lakes, coupled with expanded output from several existing projects should put global lithium production above 1.5 million tonnes of LCE by 2025. To convert the LCE numbers to pure lithium they have to be multiplied by 0.188. There is a strong argument that further out, as momentum builds up especially with the electric automotive industry, demand could outweigh supply.

Fusion does not play any role in these estimates, so what do they mean for fusion? Can its lithium requirements be met? The story on the ITER website talks about the ludicrous number of 10,000 reactors. Let us be somewhat more modest and ask the following (purely hypothetical and rather premature) question: How much lithium would be required annually if fusion were to provide 30% of electricity supply at the end of this century or early in the next century, as the EFDA Roadmap would like? It can be calculated that in that case approximately 24,000 TWh would be required from 2760 fusion power stations, each providing 1 GWe. Using the number of

500 kilos of lithium-6 per year as quoted above from the ITER website, they would consume 1380 tonnes of lithium-6, for which about 17,200 tonnes of natural, non-enriched lithium would be required annually, equivalent to about 90,000 tonnes of LCE. The competition from other uses like rechargeable batteries will be fierce, but it may be possible to meet the fusion demand.

More problematic are the much larger initial lithium loadings. The sum of the lithium inventories for all power plants would represent almost one tenth of the reserves, a quantity that would just not be available in the comparably short time that these power stations have to be built. If we also include the potential of seawater, there is enough lithium, at least theoretically, for the operation of 2760 power plants for 23 million years!

As regards beryllium the following applies. This (relatively) rare chemical element with atomic number 4 is the second lightest metal (after lithium). It is a health and safety issue for workers. The primary risk is inhalation of beryllium dust.

Globally identified resources of beryllium amount to a little over 80,000 tonnes and its extraction is a difficult process. Currently the US, China and Kazakhstan are the only three countries involved in industrial-scale extraction. The US is the world's largest beryllium producer by far, with production of 170 tonnes in 2019. Global beryllium production in 2019 totalled an estimated 260 tonnes.

In ITER, beryllium will be used as armour for the plasma-facing first-wall panels fitted inside the tokamak. For this, about 12 tonnes of beryllium will be required to cover a surface of approximately 610 m^2, which will be managed through a detailed beryllium safety programme. ITER will not have a tritium-breeding blanket, but only test mock-ups of breeding blankets. Hence for ITER there will not be a supply problem, but the beryllium burnup in the 2760 power plants envisaged above would be 524 tonnes annually, and the initial loading of 331,000 tonnes (about 120 tonnes per reactor) vastly exceeds the present estimate of resources! This makes it in any case necessary for beryllium to be recovered from waste beryllium that has become radioactive in the reaction with neutrons. Methods are being developed for this, but the above figures also show that the use of beryllium on a large scale (i.e., for more than one or two reactors) as a neutron multiplier in tritium-breeding blankets is just out of the question. There simply is not enough beryllium around.

The situation for lead is somewhat better: for the 2760 reactors mentioned above the annual burnup would be 8560 tonnes and the initial loading 11.3 million tonnes. Global lead reserves are about 85 million tonnes. Total global consumption of refined lead in 2018 amounted to 11.7 million tonnes, so if fusion comes around, that production has to be doubled to provide the initial loading, but the burnup is such that there would be sufficient lead for quite a number of years. The situation is not optimal, but better than for beryllium or lithium.

These are all rather speculative and premature calculations, and it must be said that in the past, tales of doom have been spread around about the world running out of resources. These stories never came true; resources just refuse to run out, it seems. Resources do get scarce, but then the price goes up, encouraging people to adapt by conserving it or finding cheaper substitutes. The calculations presented here in any case do show that fusion, if ever realised to the extent assumed here, might make heavy demands on the lithium supply and might come into conflict with the even heavier demand expected for the production of batteries for electric vehicles. Moreover, lithium mining has a considerable negative environmental impact which will further erode fusion's environmentally clean image. The demand for beryllium from fusion cannot be met, not even for one or two reactors. In summary, it can be stated that the shortage of these three metals (lithium, beryllium, lead) might raise problems for the fusion enterprise and that other solutions must be looked for.

Price of Electricity from Fusion

In spite of the fact that fusion has not even shown that it can generate electricity, and that such a demonstration is still at least 15–20 years away, calculations of the price of electricity generated by fusion have already been made. It will be clear that such calculations are largely meaningless, but the predilection for senseless calculations seems hard to suppress. And although such cost figures should not be taken too seriously in detail, a very precise number for the expected price of electricity from fusion is nonetheless given. In the above-mentioned European power plant study, it varied, depending on the model, from 9 eurocents per kWh to 5 eurocents per kWh, which, surprise, surprise, happens to fall squarely in the competitive range. The main point, it is said, is that the order of magnitude is not unreasonable. The conclusion of the power plant study is that "economically acceptable fusion power stations, with major safety and environmental advantages, seem to be accessible through ITER with material testing at IFMIF, and intensive development of fusion technologies." It is clear though that if the numbers had been 27 to 15 eurocents, which would still have been of the same order of magnitude, this conclusion would not have been drawn and fusion would have been decried as being uncompetitive.

Such calculations are called *ex-ante* economic analysis. Not surprisingly, the outcome of such calculations is always a competitive price for such electricity in relation to other renewable forms of energy. If that were not the case, the results would probably not be published, but sent back to the authors with the request to do better.

When discussing the cost of electricity, capital costs and running costs have to be distinguished. The capital costs are the one-time costs for putting up the power plant, viz., costs of construction, purchase of land and suchlike. They differ vastly between the various electricity generating options and are commonly expressed as overnight costs (i.e., not including any interest to be paid during construction, as if the plant were realised 'overnight') per kW of generating power of the plant to be built. As of 2019, estimated costs for some of the most common generating options are:

- gas/oil power plant—$1000/kW;
- onshore wind—$1600/kW;
- offshore wind—$6500/kW;
- solar PV—$1060–2000/kW;
- conventional hydropower—$2680/kW;
- geothermal—$2800/kW;
- coal—$3500–3800/kW;
- advanced nuclear—$6000/kW.

From this it can be seen that offshore wind and nuclear are the most expensive options, with gas/oil fired stations the cheapest, followed closely by solar and onshore wind. This reflects the huge capital costs of nuclear fission power stations (apparently taken here to be $6 billion for a 1GWe plant) and the difficulty of installing windmills at sea. It is no surprise to see solar (with its cheap solar panels) and gas/oil as the cheapest options. We don't know of course what the construction cost of a future nuclear fusion power plant will be, but if the experience with ITER is anything to go by and a 1GWe nuclear fusion power plant could be built at the same price, fusion power would be by far the most costly at $22,000/kW, more than three times the figure quoted above for advanced nuclear and in a completely different league from most solar and wind options. Running a nuclear fusion power plant must be dead cheap for it to be able to make good such high starting costs.

Running costs include the cost of fuel, maintenance costs, repair costs, wages, handling waste, etc. Fuel costs tend to be highest for oil fired generation, with coal being second and gas still cheaper. Nuclear fuel is much cheaper per kWh, and fuel is of course free for solar and wind.

For estimating the costs of electricity, a quantity called the **levelized cost of energy** (**LCOE**), or **levelized cost of electricity**, is used. It "represents the average revenue per unit of electricity generated that would be required to recover the costs of building and operating a generating plant during an assumed financial life and duty cycle." It is a measure of the average net cost of electricity generation for a generating plant over its lifetime. For technologies with no fuel costs and relatively small variable operation and management costs, such as solar and wind technologies, LCOE varies nearly in proportion to the estimated capital cost of the technology. For technologies with significant fuel cost, both fuel cost and capital cost estimates significantly affect LCOE. It is clear that most of these factors are largely unknown for nuclear fusion power plants.

Nevertheless, calculations of this kind for fusion go as far back as 1995, and have been continued ever since in all sorts of versions, even for the most outlandish types of fusion. The calculations are made on the assumptions that the fuel for fusion is inexhaustible and available at an insignificant price, nuclear safety is inherent, as well as environmental impact negligible, resulting in a possible conclusion that this provides "great scope for reducing investment cost on the basis of technological research and development with a high probability to become the cheapest and cleanest energy source from the end of this century for an unlimited time onwards" or some similar exclamation of great promise. That is quite something: "a high probability to become the cheapest and cleanest energy source (...) for an unlimited time", and that of something that does not even exist! Such calculations, based on very bold and partly dubious assumptions and extrapolations, do not warrant such conclusions. A simple example is the assumed amount of capital investment, i.e., construction costs, needed for a plant to be built early in the next century. For this a figure of just $8.5 billion is quoted (resulting in capital costs of $8500/kW, to be compared with the number for other energy generating options presented in the table above). Knowing that ITER's construction cost will far exceed $20 billion, that amount is at least a factor of 2–3 too low. Fortunately, not everybody in the fusion world jumps to this kind of outrageous conclusion.

Inherent nuclear safety is of course never mentioned for a coal plant or other type of fossil fuel burning plant, as it is obviously irrelevant for such a plant. It is standard for nuclear fusion proponents to add this feature to the advantages of nuclear fusion as they like to compare nuclear *fusion* power stations with nuclear *fission* power stations, but there is not much reason for doing so. If it were not inherently safe, nobody would even look at it. Although most currently operating nuclear power plants are not inherently safe, inherently safe or passively safe fission power stations are not impossible.

Another reason for making the comparison with nuclear fission is that fission, as we know, has insurmountable problems with nuclear waste. The nuclear waste in the case of fusion, although certainly not insignificant (see the next chapter), is a smaller problem than for fission. So, the comparison shows fusion in a favourable light, which would not be the case if the nuclear waste comparison were made with solar or wind power, as they simply have no nuclear waste problem.

The hopeless position for fusion is also illustrated by a 2005 calculation of the LCOE for fusion. In the publication all kinds of assumption are made about the 'learning factor' of fusion, and the LCOE is then compared with those for wind and solar. Solar, still underdeveloped in 2005, always comes out much more expensive than fusion, while fusion is shown to be competitive with wind, already then considered a mature technology. However, the cost development of solar, even in the few years that have since passed, has become completely different from the assumption made in that calculation. In 2014 the LCOE for solar was already at the level of around €0.20 per kWh, varying between €0.06 for the southern European countries and €0.26 for the Scandinavian countries, far below the values quoted in the paper. Solar has already advanced to a mature technology and will become even more so in the near future. Another report from 2015 states in this respect: "Solar photovoltaics (i.e., from solar panels) is already today a low-cost renewable energy technology. The cost of power from large-scale solar installations in Germany fell from over €0.40/kWh in 2005 to €0.09/kWh in 2014. Even lower prices have been reported in sunnier regions of the world, since a major share of cost components is traded on global markets." And all this while fusion can only start its 'learning curve' early in the next century, if at all, when solar and wind will be household products.

A still more recent report states: "In most parts of the world today, renewables are the lowest-cost source of new power generation. As costs for solar and wind technologies continue falling, this will become the case in even more countries" and "Onshore wind and solar PV are set by 2020 to consistently offer a less expensive source of new electricity than the least-cost fossil fuel alternative, without financial assistance". In short, the situation for fusion just seems hopeless.

Conclusion

In spite of all the difficulties we have presented here, many so-called specialists still believe that nuclear fusion will become the major energy source by the end of the 21st century, or perhaps somewhat later. They claim that there are no other alternative energy sources of equal size at the disposal of mankind even in the long run, just ignoring solar, wind and nuclear fission. Nuclear fusion, they say, apparently without blushing, has in principle been technologically mastered. The key laws of the process are understood and, although the problem has proved to be much more complicated than initially believed, it is hardly doubted that it will be solved with time. If controlled thermonuclear fusion is available, humanity will have a virtually inexhaustible energy source for thousands of years. It must be very gratifying to repeat such a sentence over and over, I suppose, since it is done so often. Optimists are hoping that after ITER a demonstration fusion power plant will be constructed by the mid-21st century, but even in that case nuclear fusion energy would not be available for the market before the end of the 21st century. So, although I do not outright disagree with the statement that "it will be solved **with** time", it is a certainty that it will not be solved **in** time. Nuclear fusion will come too late, if at all, and its case is all but hopeless.

18

Environment and Safety

There is no such thing as a perfectly clean energy option, free of any impact on the environment. Of course, when actually generating energy, solar, wind and hydropower are very clean indeed, but before they can be put into operation solar panels, windmills and dams have to be manufactured or constructed, which requires investment, materials, etc. Most simply it can be stated that they cost money and that everything that costs money will produce waste and/or cause pollution. There is no escape from this. A solar panel for instance is produced in a factory, and it uses materials of which some may be mined in an environmentally harmful manner. It has to be transported in a (polluting) ship from the production facility, most probably in China, to let's say Western Europe, where it will have a limited lifetime perhaps on a large solar park taking up land and destroying habitats. It will then be discarded, perhaps recycled, but partly dumped or removed with possible further harm to the environment. Even hydropower requires the construction of dams, which disturb ecosystems, etc.

Nuclear fusion has always been presented as the cleanest energy generating option available. In this chapter we will consider to what extent that is true and concentrate on environmental and safety issues regarding nuclear fusion power stations in their operational phase. Before they are in that phase, due to the huge construction costs involved, such power stations have already left a very considerable carbon footprint. It would be helpful to know the data of electric power consumption for, e.g., JET or ITER to be able to estimate such footprints. For ITER, at a price of more than €20 billion, it will be colossal.

L. J. Reinders, *Sun in a Bottle?... Pie in the Sky!*, https://doi.org/10.1007/978-3-030-74734-3_18

Whatever happens, nuclear fusion power generation will not be able to erase this footprint in this century by generating its 'clean' energy. For the time being all the activity and effort to bring fusion to fruition only aggravates the problem of climate change, in spite of its alleged 'great promise' to alleviate it.

That nuclear fusion power stations will not be simple facilities as regards possible environmental impact and safety is already borne out by the fact that under French law ITER is classified as a nuclear facility. This implies that a special licensing procedure has to be followed and special licenses obtained. The ITER Organization is excessively proud of the fact that ITER is the first nuclear installation to comply with the 2006 French law on Nuclear Transparency and Security and the first fusion device in history to have its safety characteristics undergo the scrutiny of a Nuclear Regulator to obtain nuclear licensing. The fact that all this is necessary shows that ITER is not just a piece of cake. Or has all this hullabaloo been raised for show? Apart from nuclear fission power stations, no other means of generating energy needs supervision by or licensing from a nuclear regulator. ITER needs such licensing as it will have enough radioactive materials on its premises to be potentially dangerous to the public and the environment. The radiotoxicity of radioactive substances (tritium and activated materials) is lower by several orders of magnitude compared to the highest-level waste from fission power plants, but the volumes of such waste for nuclear fusion power stations will be much bigger. For comparison, in its 60-year-long history up until 2019 the nuclear fission programme in the UK produced 133,000 m^3 of radioactive waste, divided into high-level waste (HLW), intermediate-level waste (ILW), low-level waste (LLW) and very-low-level waste (VLLW). Three-quarters of this waste is ILW and only 2,150 m^3 is high-level waste that needs to be stored for a very long time indeed, some say 10,000 years while others insist on a million years. High-level waste will not be produced by nuclear fusion plants, but the other categories will. And JET for instance, just a fairly basic and small-scale fusion experiment, already produced 3,000 m^3 of such waste.

Radioactivity is just one of the potential safety and environmental issues of nuclear fusion plants. Other issues include accidental releases of tritium or activated materials; thermal discharge to water or air (i.e., cooling); release of stored energy via magnet coils, atmospheric pressure on the chamber (e.g., disruptions); plant decommissioning; earthquakes, floods, storms; and aesthetic impact. We will discuss most of them in this chapter and assess their potential harmful effects.

Tritium

Tritium seems to come back in almost every chapter. Here it appears because it is radioactive, which makes it hazardous to humans, and its use in nuclear fusion power stations raises both safety and environmental issues. Tritium emits very low-energy β-radiation (i.e., electrons). It is one of the least hazardous radioactive materials, and a potential health risk only when taken inside the body, e.g., via tritiated water (HTO).

Since hydrogen isotopes are very mobile and can diffuse through materials, the containment of tritium is, however, a major safety concern. We have already noted before that for this reason scientists are in fact afraid of using tritium, and although deuterium–tritium is the preferred fusion fuel, it has so far hardly ever been used. This is one of the peculiar facts of the entire fusion enterprise. All the hullabaloo and beautiful stories about deuterium-tritium fuel are based on very scant experience indeed. In both TFTR and JET, operations with D-T plasma were severely limited, precisely because of the dangers posed by tritium. The total amount of tritium processed in TFTR was just 99 g (of which just 5 g were injected into the torus during D-T operations) and in JET 36 g. Only a small part of the injected tritium actually burns in the fusion reactions, at most a few per cent. For ITER it is less than 1% and the planned Chinese reactor CFETR calls a burnup fraction of more than 3% very challenging. The rest of the tritium escapes from the reaction region, must be recovered (in real time) from the surfaces and interiors of the reactor and its sub-systems, and re-injected ten to twenty times before it is completely burned.

A large fraction of the finally remaining tritium is extracted from the vacuum vessel using a special system and sent for recycling. This is or can be made into a well-controlled operation and will only be a possible hazard to workers and public safety in case of accidents, but it underlines what a troublesome substance tritium actually is. For ITER, a multiple-layer barrier system has been designed as protection against the spread or release of tritium into the environment. If all works well, this tritium containment system should be adequate to prevent tritium from escaping. Part of the tritium, however, remains trapped in the materials of the vacuum vessel, e.g., in the plasma facing components. In TFTR and JET about 13% and 10%, respectively, of the tritium still remained after various cleaning methods had been applied. This tritium diffuses further into cooling systems and other parts of the plant. All this has meant that only now, in 2021, for the first time since 1997, is JET preparing for a resumption of tritium operations. For ITER about 3 kg of tritium will be needed for start-up, and the total amount needed for the basic physics phase has been estimated to be in the range of 15–30 kg, so the problem for ITER will be vastly bigger than it is for JET.

In short, tritium is very unpleasant to work with. To limit the radiological risk to workers and the public, it is a requirement that the amount of tritium in each part of the plant should not exceed a predetermined maximum amount, including the amounts that are buried and continue to build up in the plasma facing components. Although the plasma burn in itself is not affected by the tritium in such components, it would be a concern if the tritium inventory allowed in the facility were exceeded, forcing a shutdown of the plant. According to the ITER website the maximum amount of tritium in the facility will not exceed 4 kg. The quantity of tritium present in the vacuum vessel will be less than or equal to 1 kg.

In spite of the experience with TFTR and JET there are still large uncertainties about tritium retention in components of the device. The use of mixed materials (carbon, tungsten, beryllium) in the plasma facing components introduces significant uncertainties in the tritium accumulation, which can only be resolved when the plant is in operation.

For ITER it is claimed that during normal operations its radiological impact on the most exposed populations will be one thousand times less than natural background radiation. There is no reason to doubt this, but accidents are also possible and non-negligible amounts of tritium could be released into the environment, e.g., from the leaking or breaking of pipes in the tritium system, and people can get exposed. When released into the air in gaseous form in such accidents, it would diffuse and disperse rather rapidly, keeping the hazard low. The situation could be more dangerous to humans when released in the form of tritiated water. Tritium in tritiated water (some of the coolant will get contaminated by tritium) always causes difficulties in nuclear installations, including equipment corrosion. In Japan the health effects of tritium are a major public concern due to the presence of tritium-contaminated water (HTO) at the Fukushima nuclear power plant, although there was no direct danger to the public.

As stated on the ITER website, even in the event of a cataclysmic breach in the tokamak, the levels of radioactivity outside the ITER enclosure would remain very low. For postulated "worst-case scenarios," such as fire in the tritium plant, the evacuation of neighbouring populations would not be necessary. In spite of this, to prevent accidental tritium releases, ITER and any other fusion facility need, just like a fission reactor, to implement a safety concept. The safety demonstration for the ITER facility, the first of its kind, still needs additional studies and all facilities created after ITER will raise even more significant nuclear safety and radiation protection issues. If not properly addressed, they could be an obstacle (in terms of delay or additional cost) or a stop (no licensing) on the way to fusion energy.

The last point to be mentioned in respect of tritium is tritiated waste. As soon as tritium is used as fuel for the fusion reaction, in-vessel components will be contaminated by tritium adsorption (adhesion to surfaces) and permeation. Components of the fuelling system and tritium plant will likewise be tritiated due to tritium permeation. High-level tritiated waste will for instance be produced from the regular replacement of plasma facing components. Furthermore, tritiated water circulating in the heat transfer systems will corrode the pipes, leading to activated corrosion products in the fluids and pipe circuits. All this waste must be detritiated, i.e., the tritium must be removed, before it can be disposed of.

The situation in ITER is still fairly simple as far as tritium is concerned. Future nuclear fusion power plants will use tritium on a continuous basis and will have to breed their own tritium, in the tritium-breeding blanket surrounding the vacuum vessel. They must do so in sufficient quantities to become self-sufficient, which will be very hard because some of the tritium will escape and cannot be recovered, as experience with JET and TFTR has shown. If only one per cent of the unburned tritium is not recovered and reinjected, even the largest surplus in the lithium-blanket regeneration process will not be able to make up for the lost tritium. The quantities of tritium in any follow-up plant will be considerably greater than in ITER. This, together with the tritium breeding, compounds the problems involved in the safe handling of tritium.

This discussion shows that nuclear fusion with deuterium-tritium as fuel is not such a clean option after all and that the use of tritium raises considerable safety and environmental issues. The tritium problem is certainly highly non-trivial. Renewables like solar and wind energy have no such problems, nor do they need licensing as a nuclear installation.

Radioactive Waste

A further important point is radioactive waste. Activation of the structure will already start with the 2.5 MeV neutrons resulting from the D-D operations in the period preceding the D-T operations, but the stream of highly energetic (14 MeV) neutrons from fusion reactions in D-T plasmas will produce huge volumes of radioactive waste as they bombard the walls of the reactor vessel and its associated components (see Chap. 11 where this was discussed as an issue when choosing the materials to be used for the plasma facing components). Damage to exposed materials from neutron radiation, which causes swelling, embrittlement and fatigue, is a long-recognized drawback of fusion energy. The total operating time with high neutron production in ITER, which equals the number of D-T shots times the pulse length of 400 s, will be too small to cause any significant damage to structural integrity, but neutron interactions will still create dangerous radioactivity in all exposed reactor components, eventually producing a staggering 30,000 tonnes of radioactive waste. Nobody in France seems to worry much about this, although France will be responsible for the decommissioning of the device. After the final shutdown, and following a five year 'deactivation' period, responsibility will be transferred to France.

For fusion power plants, which are supposed to work in steady-state, this will be a very severe problem as the volume of radioactive waste will be much larger still, much larger than for a fission power plant, but fortunately much less dangerous. The waste from nuclear fusion is classified asvery low, low, or intermediate level waste (VLLW, LLW or ILW) , but is nonetheless radioactive waste and must be treated as such. At ITER all waste materials (such as components removed by remote handling during operation) will be treated, packaged, and stored on site for the duration of the ITER experiment. A monstrous concrete cylinder 3.2 m thick, 30 m in diameter and 30 m tall, surrounding the ITER tokamak and called the bioshield, will prevent X-rays, gamma rays and stray neutrons from reaching the outside world. The reactor vessel and other components both inside the vessel and beyond, up to the bioshield, will become radioactive by activation from the neutron streams. Among other things, it implies that downtimes for maintenance and repair will be prolonged because all maintenance must be performed by remote handling equipment. For the much smaller JET project the radioactive waste volume is estimated at 3,000 cubic metres, and the decommissioning cost will exceed $300 million, according to the *Financial Times*. Those numbers will be dwarfed by ITER's 30,000 tonnes of such waste, which will in turn be dwarfed by the waste produced by power stations that operate continuously for a considerable period of time. Most of this induced radioactivity will decay in decades, but that does not make the radiation less real. ITER's Final Design Report reckons that after 100 years some 6,000 tonnes will still be dangerously radioactive and require disposal in a repository.

A recent analysis comparing total radiotoxicity of fission reactors (both currently operating and future reactors) and fusion reactors arrived at the conclusion that fusion reactors have higher radiotoxicity due to short-lived radionuclides, mainly activation products of structural materials and breeders.

After some decades of decay, the situation changes completely. For instance, after 100 years of decay, the total radiotoxicity of fusion reactors becomes 100 times lower than that of fission reactors, and after some 500 years the total radiotoxicity of fusion reactors has fallen to levels close to the natural radiotoxicity contained in fly ashes from coal-burning power plants. From this it is clear that radiotoxicity from nuclear fusion plants is nontrivial, even serious. There is, however, a high degree of uncertainty in such estimates, as components added to steel to improve its mechanical and thermodynamic properties greatly affect long-term activity. Even small levels of impurities can have significant impact on the severity of the produced radioactive waste.

The 100 years necessary for storage could possibly be reduced for future devices through the development of 'low activation' materials, like RAFM steel, which is an important part of fusion research and development today. Some call these 100 years fortunate, since they compare it with the waste from fission plants, which has to be stored for millions of years; but 100 years is still a very long time. Most people would consider 100 years (four generations) long-lasting; imagine that you still have to look after the waste produced by your grandfather or great-grandfather, on top of the waste you yourself are producing. Moreover, for waste from fission plants it is also true that, when simply left to decay radioactively for 40 years, its radioactivity has dropped by 99.9%. Radioactive waste and the accompanying radiation belong to the most important problems fusion will be faced with. Just imagine how much waste the hypothetical 2760 power plants mentioned in Chap. 17 would produce!

Scarcity of Materials

Many materials are needed for the construction of nuclear fusion power plants. In the previous chapter we have already discussed the fact that in the future difficulties might arise in connection with the availability of lithium and beryllium. Other elements that may be in short supply include helium, copper, chromium, molybdenum, nickel, niobium, tungsten and some rare-earth elements. Niobium should be avoided in structural materials exposed to neutrons, because of induced radioactivity, but is used in superconducting magnet coils. Future fusion power plants will probably be less prolific users of both niobium and helium than for instance ITER as they will not use niobium-based, helium-cooled low-temperature superconductors but high-temperature superconductors. The ITER cryogenic system requires 24 tonnes of helium, primarily for the cooling of its superconducting magnets.

Helium is a rare element and a non-renewable resource. The cumulative demand up to 2050 for cooling superconducting transmission lines is estimated at about 12 billion cubic metres. This demand could limit the helium resources available for fusion power plants (a similar situation to the one for lithium). The global production of helium in 2019 was 160 million cubic metres. Helium production is traditionally dominated by the US, a bit less than half of the total global production in 2019, 68 million cubic metres, was extracted from natural gas in the US. One of the major issues of helium production is that a large amount of helium is lost, due to the venting of helium rich gases in the natural gas industry. Fusion also produces helium (making it partly into a renewable resource) as one the products of the nuclear fusion reactions and of the tritium breeding in the blanket. It is important that this very valuable 'ash' should be collected for future use.

Niobium is most commonly used to create alloys. Even as little as 0.01% of niobium markedly improves the strength of steel. It is also used in small amounts in superconducting wires. Global production in 2019 amounted to about 75,000 tonnes, with 65,000 tonnes coming from Brazil. Demand is not expected to outstrip supply.

The world's reserves of tungsten are 3.2 million tonnes, mostly located in China (1.8 million tonnes), Canada, Russia, Vietnam and Bolivia. China dominates the market with a share of about 80%. Approximately half of the tungsten is consumed for the production of hard materials—namely tungsten carbide—with the remaining major use being in alloys and steels. Demand is rising, but supply and demand are still roughly in equilibrium. In ITER it will be used as the plasma facing material of the divertor. It is not clear what the effect on world supply will be when tungsten is applied on a large scale in fusion power stations.

In summary, shortages in materials for fusion most probably concern lithium, beryllium, and helium. Shortages of some materials might also pose problems for renewables like solar and wind for that matter. It has been estimated that, for equivalent installed capacity, solar and wind facilities require at least an order of magnitude more concrete, glass, iron, copper and aluminium than fossil fuel or nuclear energy power plants.

Plasma Disruptions and Quenching

In Chapter 10 we have already noted that disruptions and quenching are safety issues for ITER. Disruptions are expensive and dangerous. The heat can be ten times higher than the melting point of the first wall and the divertor.

A disruption occurs when an instability grows in the tokamak plasma to the point where there is a rapid loss of the stored thermal and magnetic energy. This rapid loss can also accelerate electrons to very high energy (runaway electrons). To safeguard the device against such an eventuality a special disruption mitigation system has been envisaged to protect the plasma-facing components against the resulting forces and heat, and at the same time tame the runaway electrons. See Chap. 11 for more details about disruptions and the system chosen to protect ITER. If this system does not work, it will be the end of ITER. Damage will probably be extensive and, in view of the intricacy of the device and the required fine-tuning of the components, repair will be difficult.

A safety analysis has suggested an occurrence of such an event by the failure of a penetration line, e.g., a diagnostic line of which there are many in ITER. Such failure leads to pressure build-up in the vacuum vessel due to gas (e.g., air) inflow. Plasma burning is terminated, and a disruption triggered. The vacuum vessel pressure increases until the pressures inside and outside the vessel are almost equal. The air inside the vessel is then heated by hot component surfaces, causing expansion of the vessel atmosphere and the air to flow out of the vessel.

The second problem is quenching, the situation when a superconducting magnet suddenly becomes a normal electromagnet and releases its energy. When cooled to around minus 269 °C, ITER's magnets become powerful superconductors. The electrical current surging through a superconductor encounters no electrical resistance. This allows superconducting magnets to carry the high current and produce the strong magnetic fields that are essential for ITER's experiments. Superconductivity can be maintained as long as certain threshold conditions are respected (temperature, current density, magnetic field). Outside of these boundary conditions a magnet will return to its normal resistive state and the high current will produce high heat and voltage. This transition from a superconducting to a resistive state is referred to as a quench. During a quench, temperature, voltage and mechanical stresses increase. A quench that begins in one part of a superconducting coil can propagate, causing other areas to lose their superconductivity. As this phenomenon builds up, it is essential to discharge the huge energy accumulated in the magnet to the exterior of the tokamak building.

ITER's coils contain the same energy as 10 tonnes of TNT. Quenching causes components to overheat and melt; it may even start dangerous fires. Such events may occur as the result of mechanical movements that generate heat in one part of the magnet. Variations in magnetic flux or radiation coming from the plasma can also cause quenches, as well as issues in the magnet's cryogenic coolant system. ITER is developing an early quench detection system to protect its magnets.

Magnet quenches are not expected often during the lifetime of ITER, but it is necessary to plan for them. They are actually not accidents, failures or defects, but part of the life of a superconducting magnet and the latter must be designed to withstand them. An early quench detection system must make it possible to react rapidly in order to protect the integrity of the coils, avoid unnecessary machine downtime, and discharge large amounts of stored energy to avoid damage to the vacuum vessel. There is some experience with quench detection at the Large Hadron Collider at CERN, which can be used at ITER, but to have an appropriately working system at ITER is nonetheless a tremendous challenge.

Water Usage

ITER will have a cooling water system designed to release all the heat generated from its components (nuclear and non-nuclear) into the environment by using water as coolant. The only exception is the vacuum vessel, whose heat is released to air coolers, via a separate heat transfer system. The total heat to be released to the environment is on average 500 MW during the D-T pulse, with a peak of about 1100 MW during the plasma burn.

Approximately 3 million cubic metres of water, roughly the annual usage of 300,000 people, will each year be needed during ITER's operational phase. This water will be supplied by the nearby Canal de Provence and transported through underground tunnels to the fusion installation. The water volume needed for ITER represents 1% of the total volume transported by the Canal de Provence. The combined consumption of the ITER installation and the adjacent CEA facilities remains below 5% of the total water volume transported by this canal.

One of the points of criticism is that ITER needs torrential water flows to remove heat from various parts and components of the facility. Including fusion generation, the total heat load could be as high as 1000 MW, but even with zero fusion power the reactor facility consumes huge amounts of electricity which eventually becomes heat that has to be removed. ITER will show that fusion reactors will be much greater consumers of water than any other type of power generator, because of the huge parasitic power drains that turn into additional heat that needs to be dissipated on site. (By "parasitic," we mean consuming a chunk of the very power that the reactor produces.) During fusion operations, the combined flow rate of all the cooling water will be as high as 12 cubic metres per second, or more than one third of the flow rate of the Canal de Provence. The actual demand on the Canal's water will be only a fraction of that value because ITER's power pulse will be just 400 s, with at most 20 such pulses daily, and ITER's cooling water is recirculated. But, the important point here is that, while ITER is producing nothing but neutrons and no power, its maximum coolant flow rate will still be nearly half that of a fully functioning coal-burning or nuclear plant that generates 1000 MWe of electric power. Operation of any large fusion facility such as ITER is only possible in a location such as the Cadarache region of France, where there is access to many high-power electric grids as well as a high-throughput cool water system. In past decades, the great abundance of freshwater flows and unlimited cold ocean water made it possible to realise large numbers of gigawatt-level thermoelectric power plants. In view of the decreasing availability of freshwater and even cold ocean water worldwide, the difficulty of supplying coolant water would by itself make the wide future deployment of fusion reactors impractical. It would be hard to meet the cooling water demands of the large number of nuclear fusion power plants required, e.g., the hypothetical 2760 power plants mentioned in Chap. 17.

The cooling water also contains radionuclides, because impurities (e.g., tritium) diffuse from in-vessel components and the vacuum vessel, and should be cleaned before being returned to the Canal.

Earthquakes, Floods, Storms

The ITER facility is designed to resist an earthquake of amplitude 40 times higher and with energy 250 times higher than any earthquake for which there are historical or geological records in that area of France. Although some protection against earthquakes is sensible, this precaution seems to be completely over the top and one wonders why this has been done (window-dressing?), as an earthquake would just bring the operation to a standstill and apart from the possible release of some tritium not much else can be expected. The ITER tokamak building will be made of specially reinforced concrete, and will rest upon bearing pads, or pillars, that are designed to withstand earthquakes (as used to protect other civil engineering structures such as electric power plants from the risk of earthquakes).

The risk of flooding, too, has been considered in ITER's design and Preliminary Safety Report. In the most extreme hypothetical situation—that of a cascade of dam failures north of the ITER site—more than 30 m remain between the maximum height of the water and the base of the nuclear buildings. Again a rather superfluous precaution, which has only added to the cost.

Following the Fukushima disaster in Japan in 2011, and the resulting tsunami and nuclear accident, the French government requested that the French Nuclear Safety Authority (ASN) carry out complementary safety assessments. The decision was made to assess not only nuclear power plants, as requested at the European level, but also research infrastructures such as ITER in order to examine the resistance of a facility in the face of a set of extreme situations leading to the sequential loss of lines of defence, such as very severe flooding, a severe earthquake beyond that postulated in the ITER safety case, or a combination of both. It seems to be an exercise in futility to study the effect of extreme climatic conditions such as tornados, hailstorms, etc., on ITER in the south of France where such violent events never happen. It may be that lessons can be learned from this for similar devices at other places, but for ITER, whose lifetime will only be 25 years, there are more serious matters that deserve attention.

Aesthetics

This is a very minor point, hardly worth mentioning. A nuclear fusion power plant is neither less, nor more beautiful or attractive to look at than any other power plant. Some may prefer such a localised structure over large solar or windmill parks, which when constructed on land are indeed an eyesore. It becomes a different matter if in the decommissioning phase it were decided to entomb the plant on site. Entombment seems to be the cheapest option, but leaves a carbuncle on the face of the Earth. The site will be unsuitable for other use and perhaps a hazard for the future. Mechanical disassembly and recycling is by far the preferable option but will be costly and time-consuming.

Conclusion

From the above it can be concluded that power generation from fusion is not as environmentally clean and safe as claimed. Tritium and radioactive waste pose considerable problems, vastly bigger problems than any problem posed by solar and wind energy. Disruptions and quenching must be avoided at all cost. The only thing that is perfectly clear is that, when in operation, a fusion power plant will not produce any carbon dioxide and will therefore not contribute to global warming. Its construction will, however, involve a large carbon footprint and how this will work out for maintenance and downtime periods, during which huge amounts of energy, generated perhaps by fossil-fuel burning plants, will be consumed, is at present anybody's guess. Whether the benefit of being carbon free is bigger than the disadvantages stated above is difficult to say, but should be assessed with a clear head, not by making some flippant remarks about "unlimited, inexhaustible, clean energy", certainly not when such remarks are untrue.

In this respect it is worrying that fusion websites (from official organisations) are not clear about the environmental impact of fusion. They sometimes compare fusion with coal stations when that is convenient for their purpose and it seems advantageous (and/or fashionable) to stress the point that fusion is carbon free. The fact that fusion power has clear environmental benefits compared to coal-fired power is actually completely irrelevant for the future of fusion. Coal-fired power stations are being phased out almost everywhere and will probably be all but gone before the first nuclear fusion power station is working, if that is ever going to be the case. At other times they use a comparison with fission plants since fusion power plants are inherently safe, with no possibility of a "meltdown" or "runaway reactions", no fission products and no plutonium production. Comparisons with solar and wind energy generation are seldom made, as fusion is then obviously at a disadvantage, certainly so long as steady-state power generation from fusion is still a faraway dream. Plainly wrong statements are often made. For instance, on the Fusion for Energy website it was stated that "a fusion reactor is like a gas burner with all the fuel injected being 'burnt' in the fusion reaction." This statement is actually untrue as only a small part of the fuel is burnt (see above) and worrying tritium contamination builds up in the reactor structure.

The fuel needed for fusion is normally referred to as inexhaustible and readily available from seawater, for instance: "The fuel it requires is abundant everywhere on the planet reducing the risk of any geopolitical tension; it is extracted from sea water and the crust of the Earth." That may be true for deuterium, but it is not so for tritium and the lack of tritium may very well increase geopolitical tensions. The tritium produced on Earth in two or three decades from today will not even be sufficient to operate two or three DEMO plants. Moreover, fusion will probably also lead to a scarcity of lithium and other materials and to fierce competition with demands for other uses. You will not find any information about this on the websites of the fusion organisations.

Another statement from the Fusion for Energy website is: "Fusion machines are inherently safer (*sic*) posing no risk to populations in the vicinity, generating no long-lasting waste". As we have seen the waste will be around for 100 years. This is of course short compared to the millions of years that some waste from nuclear fission power stations must be stored, but that is irrelevant. Having huge quantities of low-level radioactive waste in your backyard for 100 years is not something to look forward to. The use of the word 'safer' in the statement above without saying safer than what, shows that the statement was earlier part of a longer one. When first made, such statements were embedded in comparisons with nuclear fission stations and then of course served a purpose. Later, as in the above quotation, they start a life of their own and are routinely being made without any reference to nuclear fission, but then they are no longer true. It has just become a subtle way to mislead.

Fusion power stations are in any case less safe than wind parks at sea or solar panels on a roof. Wind and solar are undoubtedly the cleanest forms of energy, both as regards energy generation itself and the construction/manufacture of the necessary equipment. They have no safety issues to speak of either. Fusion energy will probably also beat fission in that respect, and it will win hands down in the safety compartment. As regards coal-fired power stations one wonders if fusion is cleaner than coal-burning or gas-burning power stations equipped with carbon capture installations, after all if no carbon dioxide is released into the environment there is hardly anything against burning fossil fuels, with natural gas the prime candidate. The cost of carbon capture is rapidly decreasing. Even siphoning carbon dioxide from the atmosphere seems to be within reach with the cost of pulling a ton of carbon dioxide from the atmosphere ranging between $94 and $232 (figures from 2017). Burning natural gas for electricity generation produces 0.2 kg of CO_2 per kWh. If the capture of a ton of CO_2 cost $100, it would add $0.02 to the price of 1 kWh of electricity generated by burning natural gas, adding less than $100 to the annual electricity bill of a typical Dutch family. The difference in construction costs between a coal- or gas-burning plant and a nuclear fusion plant lets you capture quite a lot of carbon before running out of funds. It is also a technology that seems to progress faster than the cumbersome route of nuclear fusion, whereby construction times are so long that it cannot keep abreast of technological developments.

19

Summary and Final Conclusion

In the preceding chapters we have journeyed from the early days of nuclear fusion just after World War II all the way to the present day. Almost from the very beginning it was realised that a configuration of magnetic fields offered the best chance of confining a plasma, with torus-like devices showing the most promise. Research was at first kept secret and only three countries, the United States, Britain and the Soviet Union, had nuclear fusion research programmes. Nonetheless, in 1955 predictions were already being made that commercial power generation from nuclear fusion would be realised within two decades. Such rash and unwarranted predictions have remained a constant factor in the fusion saga ever since.

Multiple breakthroughs have been announced over the years, but so far none has led to any really significant progress. The history of nuclear fusion until about 1970 was shaped by two pivotal moments. The first was the declassification set in motion by the second Geneva conference in 1958, the result of a lack of both success and military applications. It informed the world about the secret work carried out by the three powers mentioned above.

The second pivotal moment was the apparent triumph of the Soviet-designed tokamak in 1968, which suddenly, after a time of frustrating deadlock, seemed to provide a route forward and caused a veritable global stampede into tokamak research. Many other countries were convinced that they, too, should get involved through the fear of missing out on perhaps the greatest bonanza of all time: cheap and abundant energy. They all scrambled

© The Author(s), under exclusive license to Springer Nature
Switzerland AG 2021
L. J. Reinders, *Sun in a Bottle?... Pie in the Sky!*,
https://doi.org/10.1007/978-3-030-74734-3_19

to develop their own programmes without any specific guidance or plans, just doing what the rest had already been doing for some time.

Attempts with other designs like stellarators, mirrors and various pinches were summarily (and probably foolishly) abandoned and since roughly 1970 all have been steadfastly trotting in the same direction, ignoring failures on the way or declaring them triumphs, and just hoping for the best.

A further momentous development that took place around the same time due to the overly optimistic attitude after the 1968 tokamak result was the switch from a research-oriented approach to the goal-oriented approach of commercial power generation, i.e., from a scientific research project, fusion was suddenly turned into an engineering project. It is clear that this switch came too early. Fusion was not ready for this. The behaviour of a plasma was (and still is) insufficiently understood to embark on such ventures. It was nevertheless done, but soon ran into new problems. This time with the heating of the plasma: confinement got worse when the plasma temperature was increased by external heating. Another breakthrough was needed, which came in 1982 when the so-called high-confinement mode was (accidentally) discovered in the ASDEX tokamak in Germany. This gave a renewed impetus and perspective to tokamaks, especially the European JET and the Japanese JT-60, which came online in the 1990s. Although it was realised very early on in the game that plasmas of deuterium and tritium offered the best prospects for fusion, TFTR and JET, which operated during the 1990s, have so far been the only tokamaks that have carried out a rather limited number of experiments with such plasmas. The main reason for this is that tritium is very unpleasant to work with. They are also the only tokamaks that managed to generate some fusion power, albeit much less power than needed to run the system (a paltry 2% in the case of JET and even less for TFTR).

Then politics got involved and the design of a follow-up device became mixed up with détente politics between East and West and Gorbachev's vain attempts to save the Soviet Union. After much toing and froing, ITER came about in the form of an impossibly large international collaboration, spanning more than half the Earth, a voracious monster that gobbles up most of the world's public funds for nuclear fusion. After a long gestation period from the early 1980s, ITER has been under construction in the south of France since 2013 with first plasma expected at the very end of 2025. The project has been dogged by huge cost overruns and delays. A few years before 2040 a deuterium-tritium plasma must start to burn and prove that controlled nuclear fusion is in principle possible.

Due to the exceedingly long time of trial and error with tokamaks without any sound result to show for, doubt is now slowly creeping in as to whether standard tokamaks, like ITER and most of the other tokamaks built since the 1970s, are the right solution to the problem of generating energy from nuclear fusion. Some are putting their money on spherical tokamaks, reviving older approaches, of which the stellarator seems to stand the best chance, or relying on non-mainstream approaches that were already discarded in the past. The recent flow of venture capital has brought a return of the almost frivolous atmosphere of the 1950s when the most outlandish proposals managed to get funded.

Challenging engineering and materials issues remain to be solved before any of the plans for DEMO reactors or power plants can become a reality. It involves issues that can make or break the future of fusion. In spite of all the hullabaloo and the brave optimistic face that is being put up by the fusion proponents, nothing is certain yet, far from it. New materials have to be developed that can withstand the onslaught of the 14 MeV neutrons produced in the fusion reactions, but another major issue is the self-sufficiency of tritium that may well be insoluble. The breeding of sufficient tritium in a blanket surrounding the reaction chamber is a must if fusion with deuterium-tritium plasmas is to succeed.

Fusion is claimed to be safe and environmentally friendly, a claim that is built on shaky grounds and blatantly untrue as regards the huge volumes of radioactive waste produced by fusion reactors. If fusion ever starts to generate power in quantities that meet a significant part of the electricity needs of the world, thousands and thousands of tonnes of radioactive waste will have to be stored for up to 100 years. Since the activation of parts of the reactor by the highly energetic neutrons is an ongoing process, such waste will be produced year after year, not just at the end of the lifetime of the reactor. Fusion proponents tend to hide this issue behind the irrelevant comparison with the radioactive waste from nuclear fission plants. Indeed, the waste is less dangerous than some of the waste from nuclear fission power stations, but it is nevertheless very real and a problem the general public is probably not at all aware of. The comparison with the waste from nuclear fission stations is just an attempt to belittle the problem: "if you wish to make grey look white, put it against black."

This book's central thesis is that there is no chance whatsoever that in this century nuclear fusion will make a contribution of any significance to the carbon-free energy mix. It will therefore not play a role in the urgent decarbonization of energy production, in spite of the claims to the contrary made by the fusion community. No commercially viable nuclear fusion power station will be working in this century, and probably never will, quite a climb-down from its original boast of "too cheap to meter", and in spite of all the "breakthroughs" made along the way.

It is clear that the emperor has no clothes and never had any. This conclusion stands, even if everything goes according to plan, which it never does (just think about the rather simple mishap of a magnet shorting out with the upgrade of the NSTX in 2016, a problem that now, four years later, has still not been sorted).

The main conclusion of the book can be summarised as follows:
"If ITER fulfils all its promises, meaning that it:
shows that energy production by controlled nuclear fusion is *in principle* possible (i.e., that it succeeds in achieving $Q = 10$);
shows that tritium production in a lithium or other blanket is possible;
shows that this tritium can be collected in sufficient quantities;
shows that disruptions and instabilities can be kept at bay;
shows that the structure of the facility can withstand the onslaught of the neutrons (not such a problem in ITER, but a very real one for any follow-up device),
then a still bigger DEMO device must first be built in which all this experience will have to be incorporated, resulting in some net energy production, plus a surplus amount of tritium (self-sufficiency in tritium is a must and nobody has so far shown in any convincing way that this is possible). The DEMO must also show that the materials used, some of them still to be developed, will not be activated too easily.
Such a DEMO can be built at the earliest after 2040, will need a further twenty odd years to show its viability and resolve the problems and diffi-culties that will undoubtedly arise, before the construction of the first pilot power plant can be contemplated, constructed and set to work. Results from DEMO cannot be expected before the 2060s, after which the pilot plant will have to show that commercial energy production from fusion is indeed possible. This will take us at least into the 2080s, if not considerably later still, after which the construction of real commercial power stations may possibly get under way. Hundreds, if not thousands, of such power stations will be needed for nuclear fusion to make a sizable contribution to energy genera-tion. The scenario in Chap. 17 requires 2760 fusion power plants to provide 30% of baseload electricity by the end of this century; so, a paltry 3% would already require 276 power plants! There is simply not enough time to build a sizable number of power plants in the less than two decades left for this. The experience laid out in this book shows beyond any reasonable doubt that all

the remaining eighty years of this century will be far too short for this to happen.

Moreover, when at the end of the century we take stock of all these 'success' stories, it will turn out that a nuclear fusion power station will be hideously complex, hugely expensive and extremely unreliable, and can never compete with any of the other carbon-free or low-carbon options that will by then have reached maturity. None of the forms in which nuclear fusion is currently presented to the world will contribute to a solution of the climate change problems mankind is faced with. Then, in the 22nd century and beyond, there will be little prospect for nuclear fusion as an energy generating option. Nothing, apart from a miracle, can change this."

Although there is no possibility that energy from nuclear fusion will make a tangible contribution to electricity generation in this century, the proponents of nuclear fusion nonetheless steadfastly try to fool us into believing that this could be the case, but they are actually fooling themselves. As Richard Feynman said "the first principle is that you must not fool yourself–and you are the easiest person to fool." The history of fusion shows that the human capacity for self-deception knows no bounds.

The question is whether they have crossed the red line from foolishness to fraud. Is fusion now at a stage where they are trying to deliberately fool the entire world? Does nuclear fusion deserve a place in the gallery of infinite-energy fraudsters and pranksters portrayed so eloquently and humorously, but also with repressed anger by Robert Park in his book *Voodoo Science*? The ones Park describes are rather innocent, misguided individuals who were at first genuinely convinced that they had hit a hitherto unknown jackpot. ITER and everything around it is much more serious and much more costly. If you are watching the videos of Europe's Fusion for Energy Organisation on YouTube[1] with the solemn voice of the speaker spelling out the huge numbers involved in the ITER exercise, the thousands of people working on its 24,000 tons, three times the Eifel tower, the many references to the power of the Sun, while giving you a close look at the advanced scientific process of … pouring concrete, and a tokamak building which like a true Tower of Babel is partly shrouded in swirling clouds, against the backdrop of slowly swelling almost liturgical music, you get the impression of attending a religious service, where man is in awe of and worshipping his own creation, like the Jewish people on their flight from Egypt worshipping the golden calf in the desert in Moses's absence on Mount Sinai. Only now no Moses will be coming down to command them to mend their ways. The video is clearly

---

[1] A beautiful example is https://www.youtube.com/watch?v=zqEkkN0f59E, accessed August 2020.

meant to impress the public and has hardly any further content, just empty rhetoric, enumerating numbers to show off the massiveness of the project, the larger the better, gigantomania at its peak; numbers the ITER Organization foolishly seems to be excessively proud of. The connection with science is lost. Here, too, it is apt to quote Richard Feynman who said that "for a successful technology, reality must take precedence over public relations, for Nature cannot be fooled".

In the early 1970s the original research programs, oriented towards plasma physics and at understanding its intricacies, were hijacked by pragmatists with too little understanding of science and of how science works and were forcefully made to enter upon a vain march towards a perceived cornucopia of unlimited, cheap energy, heaping hype upon hype, declaring breakthrough after breakthrough without encountering much criticism or in-depth questioning from the media or from fellow scientists (while those who vented criticism were beaten into silence). By now it has become a juggernaut that thunders on and will only be stopped when ITER makes a hard landing around 2035–2040. It has also captured a (too) large part of the science budgets in many countries, especially in the European Union with its huge contribution to ITER, leaving less and less for others. Many people and industries are partly dependent on the fusion fleshpots, will resist any change and will continue to argue that commercial energy production by nuclear fusion is just around the corner. Well, it isn't and probably never will be.

Websites

ITER	https://www.iter.org/.
ARIES	http://aries.ucsd.edu/; archived at http://qed fusion.org/bib.shtml.
Fusion Power Associates	https://fusionpower.org/.
All the worlds tokamaks	https://alltheworldstokamaks.wordpress.com.
All the worlds tokamaks	http://tokamak.info/.
Tihiro Ohkawa	https://www.fusion-holy-grail.net/.
Lops Alamos National Lab	https://www.lanl.gov/.
Fire project archive	http://fire.pppl.gov/.
Princeton Plasma Physics Laboratory	https://www.pppl.gov/.
Max Planck Society	https://pure.mpg.de/.
Max Planck Institute for Plasma Physics	https://www.ipp.mpg.de/.
French Institute for Magnetic Fusion Research	http://www-fusion-magnetique.cea.fr/.
WEST tokamak	http://irfm.cea.fr/en/west/index.php.
Italian Fusion Research	http://www.fusione.enea.it/.
EUROfusion	https://www.euro-fusion.org/.
Fusion for Energy (F4E)	https://fusionforenergy.europa.eu/.

© The Editor(s) (if applicable) and The Author(s), under exclusive license to Springer Nature Switzerland AG 2021
L. J. Reinders, *Sun in a Bottle?... Pie in the Sky!*,
https://doi.org/10.1007/978-3-030-74734-3

ENEA	http://www.frascati.enea.it/.
Korea Institute of Fusion Energy	https://www.nfri.re.kr/eng/index.
Culham Centre for Fusion Energy	http://www.ccfe.ac.uk/.
JT-60SA	http://www.jt60sa.org/.
The European Fusion Education Network	http://www.fusenet.eu/.
New Energy Times	https://news.newenergytimes.net/.
International Energy Agency	https://www.iea.org/.
Industry tap into news	http://www.industrytap.com/iter-will-never-lead-commercial-viability/32484.
NSTX at PPPL	https://nstx.pppl.gov/.
Japanese Fusion Tokyo	http://fusion.k.u-tokyo.ac.jp/.
University of Wisconsin	https://news.wisc.edu/federally-funded-upgrade-reenergizes-fusion-experiment/; https://news.wisc.edu/.
Pulsar Fusion	https://www.pulsarfusion.com/.
Tokamak Energy	https://www.tokamakenergy.co.uk/.
Center for Plasma-Material Interaction	https://cpmi.illinois.edu/2016/04/26/hidra-hybrid-illinois-device-for-research-and-applications/.
CTFusion	https://ctfusion.net/.
Plasma Universe	www.plasma-universe.com/.
LLPFusion	https://lppfusion.com/.
Fusor forum	http://fusor.net/.
Farnsworth website	http://farnovision.com/.
How to make a fusor	https://makezine.com/projects/make-36-boards/nuclear-fusor/.
General Fusion	https://generalfusion.com/.
Sandia National Laboratory	https://www.sandia.gov/z-machine/index.html.
MIT Plasma Science and Fusion Center	https://www.psfc.mit.edu/.
Commonwealth Fusion Systems	https://cfs.energy/.
Fusion Industry Association	https://www.fusionindustryassociation.org/.
American Fusion Project	http://americanfusionproject.org/.

American Security Project	https://www.americansecurityproject.org/.
FLC Business	https://federallabs.org/.
Helion Energy	https://www.helionenergy.com/.
ARPA-E	https://arpa-e.energy.gov/.
Compact Fusion Systems	https://www.compactfusionsystems.com/.
Horne Technologies	https://www.hornetechnologies.com/.
Lockheed Martin	https://www.lockheedmartin.com/en-us/products/compact-fusion.html.
TAE Technologies	https://tae.com/company/.
TerraPower	https://terrapower.com/.
Next Big Future	https://www.nextbigfuture.com/2019/01/nuclear-fusion-commercialization-race.html#more-153607.
HB11	https://www.hb11.energy/.
First Light Fusion	https://firstlightfusion.com/.
Proton Scientific	http://protonscientific.com/.
Zap Energy	https://www.zapenergyinc.com/.
AGNI	https://www.agnifusion.org/.
Type One Energy	https://www.typeoneenergy.com/.
Hyper Jet Fusion	http://hyperjetfusion.com/.
Helicity Space	https://www.helicityspace.com/.
Renaissance Fusion	https://stellarator.energy/.
INFUSE	https://infuse.ornl.gov/.
International Fusion Energy Research Centre	https://www.iferc.org/.
Japanese National Institutes	http://www.fusion.qst.go.jp/.
Fusion Technology Institute	http://fti.neep.wisc.edu/.

Literature

Many books have been written on nuclear fusion. I have benefited a lot from those mentioned below, in addition to a large number of articles in the scientific literature. For a complete list of references, please consult the extended version of this book *The Fairy Tale of Nuclear Fusion,* published by Springer Nature in 2021.

L. Badash, *Scientists and the Development of Nuclear Weapons* (Humanity Books, Amherst, 1995).

Dipak Basu (ed.), *Dictionary of Material Science and High Energy Physics* (CRC Press, 2001).

M.G. Bell, *The Tokamak Fusion Test Reactor* in: George Neilson (ed.), *Magnetic Fusion Energy: From Experiment to Power Plant* (Elsevier, 2016).

P.M. Bellan, *Spheromaks: A Practical Application Of Magnetohydrodynamic Dynamos And Plasma Self-organization* (Imperial College Press, 2000).

Amasa S. Bishop, *Project Sherwood* (Addison-Wesley, 1958).

J.L. Bobin, *Controlled Thermonuclear Fusion* (World Scientific, 2014).

Susan Boenke, *Entstehung und Entwicklung des Max-Planck-Instituts für Plasmaphysik 1955–1971* (Campus Verlag, 1991).

C.M. Braams and P.E. Stott, *Nuclear Fusion: Half a Century of Magnetic Confinement Fusion Research* (Taylor & Francis, New York, 2002).

Joan Lisa Bromberg, *Fusion–Science, Politics, and the Invention of a New Energy Source* (MIT Press, 1982).

© The Editor(s) (if applicable) and The Author(s), under exclusive license to Springer Nature Switzerland AG 2021
L. J. Reinders, *Sun in a Bottle?... Pie in the Sky!*,
https://doi.org/10.1007/978-3-030-74734-3

B. Brunelli and G.G. Leotta (eds.), *Unconventional Approaches to Fusion* (Plenum Press, New York, 1982).

B. Brunelli and H. Knoepfel (eds.), *Safety, Environmental Impact, and Economic Prospects of Nuclear Fusion* (Plenum Press, New York, 1990).

Francis F. Chen, *An Indispensable Truth* (Springer, 2011).

Francis F. Chen, *Introduction to Plasma Physics and Controlled Fusion* (Springer, 2016).

Michel Claessens, *ITER: The Giant Fusion Reactor* (Springer, 2019).

Daniel Clery, *A Piece of the Sun* (Overlook Duckworth, New York, 2013).

Barbara Curli, *Italy, Euratom and Early Research on Controlled Thermonuclear Fusion (1957–1962)* in: Elisabetta Bini, Igor Londero (eds.), *Nuclear Italy. An International History of Italian Nuclear Policies during the Cold War* (EUT Edizioni Università di Trieste, 2017).

P.A. Davidson, *An Introduction to Magnetohydrodynamics* (Cambridge University Press, 2010).

Stephen O. Dean, *Search for the Ultimate Energy Source: A History of the U.S. Fusion energy Program* (Springer, 2013).

R. Dendy (ed.), *Plasma Physics: an Introductory Course* (Cambridge University Press, 1993).

T. Dolan, *Fusion Research* (Pergamon Press, 2000).

T. Dolan (ed.), *Magnetic Fusion Technology* (Springer, 2013).

M. Eckert, *Vom 'Matterhorn' zum 'Wendelstein': Internationale Anstöße zur nationale Großforschung in der Kernfusion*, in: M. Eckert und M. Osietzki, *Wissenschaft für Macht und Markt* (Beck, München, 1989).

T. Kenneth Fowler, *The Fusion Quest* (The Johns Hopkins University Press, 1997).

Jeffrey Friedberg, *Plasma Physics and Fusion Energy* (Cambridge University Press, 2007).

G. Gamow, *My World Line* (New York, 1970).

Lucie Green, *15 Million Degrees: A Journey to the Centre of the Sun* (Penguin Books, 2016).

T.A. Heppenheimer, *The Man-Made Sun: The Quest for Fusion Power* (Omni Press, Boston, 1984).

Robin Herman, *Fusion–The search for endless energy* (Cambridge University Press, 1990).

R.L. Hirsch, R.H. Bedzek and R.M. Wendling, *The Impending World Energy Mess* (Apogee Prime, Ontario, 2010).

The JET Project: Design Proposal for the JOINT European Torus (Commission of the European Communities, 1976), EUR-JET-R5.

Paul R. Josephson, *The Red Atom, Russia's Nuclear Power Program from Stalin to Today* (Freeman and Company, New York, 2000).

Helge Kragh, *Quantum Generations* (Princeton, 1999).

M.A. Leontovich (ed.), *Plasma Physics and the Problem of Controlled Thermonuclear Reactions* (Pergamon Press, 1959), four volumes.

F. Mannone (ed.), *Safety in Tritium Handling Technology* (Kluwer Academic Publishers, 1993).

G. McCracken and P. Stott, *Fusion: the Energy of the Universe* (Academic Press, 2013).

S.V. Mirnov, *Energiya iz vody* (MIFI, Moscow, 2007).

K. Miyamoto, *Fundamentals of Plasma Physics and Controlled Fusion* (National Institute for Fusion Science, Japan, 2011).

R.L. Murray and K.E. Holbert, *Nuclear Energy –An Introduction to the Concepts, Systems and Applications of Nuclear Processes* (Elsevier, 2020).

National Academy of Sciences, *Final Report of the Committee on a Strategic Plan for U.S. Burning Plasma Research* (The National Academies Press, 2018).

Robert L. Park, *Voodoo Science: The Road from Foolishness to Fraud* (Oxford University Press, 2000).

J.M. Perlado and J. Sanz, *Irradiation Effects and Activation in Structural Material*, in: G.J. Velarde, Y. Ronen and J.M. Martinez-Val (eds.), *Nuclear Fusion by Inertial Confinement: A Comprehensive Treatise* (CRC Press, 1992).

J.A. Phillips, *Magnetic Fusion* (Los Alamos Science, 1983).

Bruce Cameron Read, *The History and Science of the Manhattan Project* (Springer, 2014).

R.Z. Sagdeev, *The Making of a Soviet Scientist* (John Wiley, 1994).

A. Sakharov, *Memoirs* (Vintage Books, New York, 1990).

Paul Schatzkin, *The Boy Who Invented Television* (TeamCom Books, 2002).

Gino Segrè and Bettina Hoerlin, *The Pope of Physics: Enrico Fermi and the Birth of the Atomic Age* (New York, 2016).

Charles Seife, *Sun in a Bottle: The Strange History of Fusion and the Science of Wishful Thinking* (Penguin Books, 2008).

E.N. Shaw, *Europe's Experiment in Fusion–The JET Joint Undertaking* (North-Holland, 1990).

V. Sivaram, *Taming the Sun–Innovations to Harness Solar Energy and Power the Planet* (MIT Press, 2018).

Weston M. Stacey, *Fusion–An Introduction to the Physics and Technology of Magnetic Confinement Fusion* (Wiley, 2009).

Weston M. Stacey, *The Quest for a Fusion Reactor: An Insider's Account of the INTOR Workshop* (Oxford University Press, 2010).

S. Tosti and N. Ghirelli (eds.), *Tritium in Fusion* (Nova Science Publishers, 2013).

W. van den Daele, W. Krohn und P. Weingart, *Geplante Forschung* (Suhrkamp, Frankfurt, 1979).

G.S. Voronov, *Storming the Fortress of Fusion* (MIR, Moscow, 1988).

J. Wesson, *The Science of JET*, JET–R(99)13 (1999).

R.B. White, *Theory of Tokamak Plasmas* (North-Holland, Amsterdam, 1989).

Hans Wilhelmsson, *Fusion –A Voyage Through the Plasma Universe* (IOP Publishing, Bristol, 2000).

Index

© The Editor(s) (if applicable) and The Author(s), under exclusive license to Springer Nature Switzerland AG 2021
L. J. Reinders, *Sun in a Bottle?... Pie in the Sky!*,
https://doi.org/10.1007/978-3-030-74734-3

Printed in the United States
by Baker & Taylor Publisher Services